Astrophysics and Space Science Proceedings

Volume 55

More information about this series at http://www.springer.com/series/7395

Christophe Sauty
Editor

JET Simulations, Experiments, and Theory

Ten Years After JETSET. What Is Next?

Editor
Christophe Sauty
Laboratoire Univers et Théories
Observatoire de Paris
Meudon, France

ISSN 1570-6591 ISSN 1570-6605 (electronic)
Astrophysics and Space Science Proceedings
ISBN 978-3-030-14130-1 ISBN 978-3-030-14128-8 (eBook)
https://doi.org/10.1007/978-3-030-14128-8

This Springer imprint is published by the registered company Springer Nature Switzerland AG.
The registered company address is: Gewerbestrasse 11, 6330 Cham, Switzerland

In memory of Jean Heyvaerts

Preface

In 2008, the European FP6 JETSET project ended. JETSET, standing for Jet Simulations, Experiments and Theory, was a joint research-training network funded by the European Union, involving 10 institutions, approximately 100 scientists, and 18 directly employed predoctoral and postdoctoral researchers, all working on the theme of jets from young stars. The present proceedings are a collection of contributions of the new results produced from those groups, since the end of the JETSET project. Some of the former JETSET students are now in the academic world, and the subject of astrophysical jets has never been more alive than it is today. Considerable progress has taken place in the field of jets from young stars in the past 10 years from the theoretical, numerical, observational, and experimental points of view. All these are opening new challenges in the field of jets from young stars. What should we expect from the development of new instruments and new numerical codes in the near future? What is the impact of the study of jets from young stars on studies of other jets, in particular those for relativistic jets? As a matter of fact, all those questions we tried to answer here show that it is the right time to plan for a new network. As a conclusion, we shall also highlight the essential and inspiring role of Kanaris Tsinganos in the field of MHD astrophysics and jets. This is also an occasion to celebrate his leading role in the JETSET network.

The goal of this book is to:

- offer an up-to-date review of what has been done in the field of young stellar jets between 2008 and 2018 in Europe, as a follow-up of the FP6 European program "JETSET,"
- provide a snapshot of what are the next theoretical, experimental, observational, and computational challenges to be addressed in the field of young stellar jets,
- discuss the input of new instruments and new numerical codes for the future,
- consider the impact of the study of jets of young stellar objects on other jet studies, in particular those for relativistic jets.

Meudon, France
May 2018

Christophe Sauty

Acknowledgments

This conference and those proceedings would never have come to reality without the inspiration and support of Kanaris Tsinganos, Professor at National and Kapodistrian University of Athens.

The editor also thanks the Observatoire de Paris and the Centre International d'Ateliers Scientifiques for their financial and logistic support as well as the Laboratoire Univers et Théories (LUTH) and its administrative staff, without which this conference would not have been possible.

Contents

List of Contributors

R. M. G. de Albuquerque Instituto de Astrofísica e Ciências do Espaço, Universidade do Porto, CAUP, Porto, Portugal

Departamento de Física e Astronomia, Faculdade de Ciências, Universidade do Porto, Porto, Portugal

Laboratoire Univers et Théories, Observatoire de Paris, UMR 8102 du CNRS, Université Paris Diderot, Meudon, France

J. M. Alcalá INAF – Osservatorio Astronomico di Capodimonte, Napoli, Italy

S. Antoniucci INAF – Osservatorio Astronomico di Roma, Monte Porzio Catone, Italy

C. Argiroffi Dip. di Fisica e Chimica, Università di Palermo, Palermo, Italy

INAF – Osservatorio Astronomico di Palermo, Palermo, Italy

F. Bacciotti Istituto Nazionale di Astrofisica – Osservatorio Astrofisico di Arcetri, Firenze, Italy

E. Bianchi Institut de Planétologie et d'Astrophysique de Grenoble (IPAG), Université Grenoble Alpes, Grenoble Cédex 9, France

M. Benisty Université Grenoble Alpes, CNRS, IPAG, Grenoble, France

K. Biazzo INAF-Osservatorio Astrofisico di Catania, Catania, Italy

G. Bodo INAF Osservatorio Astrofisico di Torino, Pino Torinese, Italy

R. Bonito INAF – Osservatorio Astronomico di Palermo, Palermo, Italy

J. Brand Italian ARC, INAF-IRA, Bologna, Italy

W. Brandner Max Planck Institute für Astronomy, Heidelberg, Germany

A. Caratti o Garatti Astronomy & Astrophysics Section, Dublin Institute for Advanced Studies, Dublin 2, Ireland

S. Casu INAF-Osservatorio Astronomico di Cagliari, Selargius, Italy

F. Cattaneo Department of Astronomy, University of Chicago, Chicago, IL, USA

V. Cayatte Laboratoire Univers et Théories, Observatoire de Paris-PSL, UMR 8102 du CNRS, Université Paris Diderot, Meudon, France

L. Chantry Laboratoire Univers et Théories, Observatoire de Paris-PSL, UMR 8102 du CNRS, Université Paris Diderot, Meudon, France

A. Ciardi Sorbonne Universités, Observatoire de Paris, PSL Research University, CNRS, LERMA, Paris, France

C. Codella Istituto Nazionale di Astrofisica – Osservatorio Astrofisico di Arcetri, Firenze, Italy

D. Coffey School of Physics, University College Dublin, Belfield, Dublin 4, Ireland

S. Colombo INAF-Osservatorio Astronomico di Palermo Palermo, Italy

LERMA, Sorbonne Université, Observatoire de Paris, Université PSL, CNRS, Paris, France

Dipartimento di Fisica & Chimica, Universitá degli Studi di Palermo, Palermo, Italy

Universidad de Las Palmas de Gran Canaria, Las Palmas, Spain

M. Damoulakis National and Kapodistrian University of Athens, Department of Physics, Panepistimiopolis Zografos, Athens Greece

O. Dionatos Institute for Astronomy (IfA), University of Vienna, Vienna, Austria

C. Dougados Université Grenoble Alpes, CNRS, IPAG, Grenoble, France

T. Downes Department of Mathematical Sciences, DCU Glasnevin Campus, Dublin, Ireland

J. Eislöffel Thüringer Landessternwarte Tautenburg, Tautenburg, Germany

F. Favre Istituto Nazionale di Astrofisica – Osservatorio Astrofisico di Arcetri, Firenze, Italy

D. Fedele INAF – Osservatorio Astrofisico di Arcetri, Firenze, Italy

C. Fendt MPI for Astronomy, Heidelberg, Germany

D. Froebrich School of Physical Sciences, Ingram Building, University of Kent, Canterbury, UK

J. Fuchs LULI CNRS, Ecole Polytechnique, Sorbonne Universités, CEA, UPME, Palaiseau cedex, France

D. Galli Istituto Nazionale di Astrofisica – Osservatorio Astrofisico di Arcetri, Firenze, Italy

J. F. Gameiro Instituto de Astrofísica e Ciências do Espaço, Universidade do Porto, CAUP, Porto, Portugal

Faculdade de Ciências, Departamento de Física e Astronomia, Universidade do Porto, Porto, Portugal

R. Garcia Lopez Astronomy & Astrophysics Section, Dublin Institute for Advanced Studies, Dublin 2, Ireland

A. Giannetti Italian ARC, INAF-IRA, Bologna, Italy

T. Giannini INAF – Osservatorio Astronomico di Roma, Monte Porzio Catone, Italy

J.M. Girart Institut de Ciències de l'Espai (ICE, CSIC), Can Magrans, s/n, Cerdanyola del Vallès, Spain

A. I. Gómez de Castro Facultad de Ciencias Matemáticas, Universidad Complutense de Madrid, Madrid, Spain

M. González Paris Diderot University, AIM, CEA, CNRS, Paris, France

O. Hervet Department of Physics, Santa Cruz Institute for Particle Physics, University of California at Santa Cruz, Santa Cruz, CA USA

I. Hubeny Steward Observatory, University of Arizona, Tucson, AZ, USA

L. Ibgui LERMA, Sorbonne Université, Observatoire de Paris, Université PSL, CNRS, Paris, France

R. Keppens KU Leuven, Leuven, Belgium

L. Labadie I. Physikalisches Institut, Universität zu Köln, Köln, Germany

T. Lanz Observatoire de la Côte d'Azur, Nice, France

S. Leurini INAF-Osservatorio Astronomico di Cagliari, Selargius, Italy

J. J. G. Lima Instituto de Astrofísica e Ciências do Espaço, Universidade do Porto, CAUP, Porto, Portugal

Faculdade de Ciências, Departamento de Física e Astronomia, Universidade do Porto, Porto, Portugal

F. Lopez-Martinez Instituto de Astrofísica e Ciências do Espaço, Universidade do Porto, CAUP, Porto, Portugal

C. F. Manara ESO Headquarters, Garching bei München, Germany

F. Massi INAF-Osservatorio Astrofisico di Arcetri, Firenze, Italy

T. Matsakos LERMA, Sorbonne Université, Observatoire de Paris, Université PSL, CNRS, Paris, France

Z. Meliani LUTH, Observatoire de Paris, CNRS UMR 8102, Université Paris-Diderot, Meudon, France

M. Miceli Dip. di Fisica e Chimica, Università di Palermo, Palermo, Italy
INAF – Osservatorio Astronomico di Palermo, Palermo, Italy

D. Millas KU Leuven, Leuven, Belgium

A. Mignone Dipartimento di Fisica, Universitá di Torino, Torino, Italy

L. Moscadelli INAF-Osservatorio Astrofisico di Arcetri, Firenze, Italy

A. Natta DIAS/School of Cosmic Physics, Dublin Institute for Advanced Studies, Astronomy and Astrophysics Section, Dublin, Ireland

B. Nisini INAF – Osservatorio Astronomico di Roma, Monte Porzio Catone, Italy

S. Orlando INAF-Osservatorio Astronomico di Palermo, Palermo, Italy

M. Padovani Istituto Nazionale di Astrofisica – Osservatorio Astrofisico di Arcetri, Firenze, Italy

R. Paladino Istituto Nazionale di Astrofisica – Istituto di Radioastronomia Bologna, Italy

M. Pedani INAF – Fundación Galileo Galilei, Breña Baja, Spain

G. Peres Dip. di Fisica e Chimica, Università di Palermo, Palermo, Italy
INAF – Osservatorio Astronomico di Palermo, Palermo, Italy

K. Perraut Université Grenoble Alpes, CNRS, IPAG, Grenoble, France

L. Podio Istituto Nazionale di Astrofisica – Osservatorio Astrofisico di Arcetri, Firenze, Italy

O. Porth Goethe-Universität Frankfurt am Main, Frankfurt am Main, Germany

F. Reale Dip. di Fisica e Chimica, Università di Palermo, Palermo, Italy
INAF – Osservatorio Astronomico di Palermo, Palermo, Italy

T. Ray Astronomy & Astrophysics Section, Dublin Institute for Advanced Studies, Dublin 2, Ireland

R. Rodriguez Universidad de Las Palmas de Gran Canaria, Las Palmas, Spain

P. Rossi INAF Osservatorio Astrofisico di Torino, Pino Torinese, Italy

L. de Sá LERMA, Sorbonne Université, Observatoire de Paris, Université PSL, CNRS, Paris, France
CEA/DSM/IRFU/SAp-AIM, CEA Saclay, CNRS, Gif-sur-Yvette, France

A. Sanna Max-Planck-Institut für Radioastronomie, Bonn, Germany

N. Sanna INAF-Osservatorio Astrofisico di Arcetri, Firenze, Italy

C. Sauty Laboratoire Univers et Théories, Observatoire de Paris, UMR 8102 du CNRS, Université Paris Diderot, Meudon, France

C. Stehlé LERMA, Sorbonne Université, Observatoire de Paris, Université PSL, CNRS, Paris, France

O. Tesileanu Extreme Light Infrastructure-Nuclear Physics (ELI-NP), "Horia Hulubei" National R&D Institute for Physics and Nuclear Engineering (IFIN-HH), Magurele, Romania

L. Testi European Southern Observatory, Garching bei München, Germany

A. Tritsis Research School of Astronomy and Astrophysics, Mount Stromlo Observatory, Canberra, ACT, Australia

K. Tsinganos National and Kapodistrian University of Athens, Section of Astrophysics, Astronomy and Mechanics, Department of Physics, Panepistimiopolis Zografos, Athens, Greece

National Observatory of Athens, Lofos Nymphon, Athens, Greece

S. Ulmer-Moll Instituto de Astrofísica e Ciências do Espaço, Universidade do Porto, CAUP, Porto, Portugal

Faculdade de Ciências, Departamento de Física e Astronomia, Universidade do Porto, Porto, Portugal

S. Ustamujic Facultad de Ciencias Matemáticas, Universidad Complutense de Madrid, Madrid, Spain

N. Vlahakis National and Kapodistrian University of Athens, Section of Astrophysics, Astronomy and Mechanics, Department of Physics, Panepistimiopolis Zografos, Athens Greece

List of Participants

R. M. G. de Albuquerque Raquel.Albuquerque@astro.up.pt

F. Bacciotti fran@arcetri.astro.it

G. Bodo gianluigi.bodo@inaf.it

R. Bonito rosaria.bonito@inaf.it

S. Bouquet serge.bouquet@cea.fr

S. Cabrit sylvie.cabrit@obspm.fr

V. Cayatte Veronique.Cayatte@obspm.fr

L. Chantry loic.chantry@obspm.fr

C. Codella codella@arcetri.inaf.it

D. Coffey deirdre.coffey@ucd.ie

S. Colombo salvatore.colombo@obspm.fr

O. Dionatos odysseas.dionatos@univie.ac.at

C. Dougados dougadoc@univ-grenoble-alpes.fr

T. Downes turlough.downes@dcu.ie

J. Eislöffel jochen@tls-tautenburg.de

C. Fendt fendt@mpia.de

D. Froebrich D.Froebrich@kent.ac.uk

R. Garcia Lopez rgarcia@cp.dias.ie

A. I. Gómez de Castro anai_gomez@mat.ucm.es

L. Ibgui laurent.ibgui@obspm.fr

R. Keppens rony.keppens@kuleuven.be

J. J. G. Lima jlima@astro.up.pt

F. Lopez-Martinez Fatima.Lopez@astro.up.pt

S. Massaglia silvano.massaglia@ph.unito.it

F. Massi fmassi@arcetri.astro.it

Z. Meliani zakaria.meliani@obspm.fr

C. Michaut claire.michaut@obspm.fr

D. Millas dimitrios.millas@kuleuven.be

B. Nisini brunella.nisini@inaf.it

S. Orlando orlando@astropa.inaf.it

L. Podio lpodio@arcetri.inaf.it

T. Ray tr@cp.dias.ie

P. Rossi paola.rossi@inaf.it

L. de Sá lionel.desa@obspm.fr

C. Sauty csauty@obspm.fr

C. Stehlé chantal.stehle@obspm.fr

B. Tabone benoit.tabone@obspm.fr

O. Tesileanu ovidiu.tesileanu@eli-np.ro

A. Tritsis Aris.Tritis@anu.edu.au

K. Tsinganos tsingan@phys.uoa.gr

S. Ustamujic sustamuj@ucm.es

N. Vlahakis vlahakis@phys.uoa.gr

C. Zanni zanni@oato.inaf.it

Part I
Theory and Models

Magnetorotational Turbulence, Dynamo Action and Transport in Convective Disks

G. Bodo, F. Cattaneo, A. Mignone, and P. Rossi

1 Introduction

The turbulence driven by the magneto-rotational instability (MRI) is thought to be at the origin of the enhanced transport of angular momentum in hot accretion disks [1]. The magnetic fields necessary for the MRI to develop can either be externally imposed or internally generated within the disk. In this latter case, the magneto-rotational turbulence also acts as a dynamo. Consequently, much effort has been devoted to the study of dynamo action within accretion disks [2–5]. Studies that incorporate the effects of the vertical stratification within the disk have shown that the operation of the dynamo depends crucially on how the disk is stratified [6–11]. Many of these studies have considered an isothermal stratification, in [10] we have instead considered the effects of turbulent heating on the disk vertical structure and have examined a case with a perfect gas equation of state and a finite thermal diffusivity (see also [12]). In this case two regimes of operation, with different dynamo properties, are possible, depending on the efficiency of thermal conduction. If thermal conduction is efficient, the behaviour is similar to the isothermal case, if, on the contrary, thermal conduction is inefficient, the disk becomes unstable

G. Bodo (✉) · P. Rossi
INAF Osservatorio Astrofisico di Torino, Pino Torinese, Italy
e-mail: gianluigi.bodo@inaf.it; paola.rossi@inaf.it

F. Cattaneo
Department of Astronomy, University of Chicago, Chicago, IL, USA
e-mail: cattaneo@oddjob.uchicago.edu

A. Mignone
Dipartimento di Fisica, Universitá di Torino, Torino, Italy
e-mail: mignone@ph.unito.it

C. Sauty (ed.), *JET Simulations, Experiments, and Theory*,
Astrophysics and Space Science Proceedings 55,
https://doi.org/10.1007/978-3-030-14128-8_1

to overturning motions (convection) and the dynamo appears to be much more efficient.

Having obtained these fully convective solutions by numerical simulations performed in the shearing box approximation, In [13] we presented a procedure for piecing together the local solutions at different radii to obtain a global solution. The final result gives the complete structure of the disk as a function of the mass of the central object and of the mass accretion rate, once the physical properties of the accreting material are specified. From this solution it is possible to check a posteriori the consistency of our assumptions and to determine the disk regions where convection occurs.

In the next section we will describe the shearing box numerical solutions for the fully convective state, in Sect. 3 we will outline the procedure for obtaining global solutions and finally in Sect. 4 we will draw our conclusions.

2 Numerical Solutions

We performed three-dimensional, numerical simulations of a perfect gas with thermal conduction in a shearing box with vertical gravity, with the PLUTO code [14]. We start our simulations from a state with a uniform shear flow, and density and pressure distributions that satisfy vertical hydrostatic balance with constant temperature. With the development of MRI driven turbulence, the disk structure will be driven away from the initial state by the energy input from dissipative processes. The temperature will increase in the equatorial region and a thermal gradient will be progressively established between this region and the disk upper and lower boundaries. The temperature increase will continue until a new equilibrium is reached where the energy input is balanced by losses at the boundaries. This new equilibrium is determined self-consistently by the heating due to MRI driven turbulence. If thermal diffusivity is large, thermal conduction can easily transport the generated heat along shallow gradients. On the contrary, if thermal diffusivity is small, the gradients required to carry the generated heat become very steep and unstable to overturning motions. In this last case it will be convection that will carry the generated heat to the surface layers of the disk. In Fig. 1 we show the temperature averaged on horizontal planes as a function of the vertical coordinate, for different values of the thermal diffusivity. On the right we show, instead, the conductive (solid) and convective (dashed) heat fluxes for two values of the thermal diffusivity. We see that decreasing the diffusivity, the temperature gradient progressively increases, while at high diffusivity the flux is mainly conductive and at low diffusivity, mainly convective.

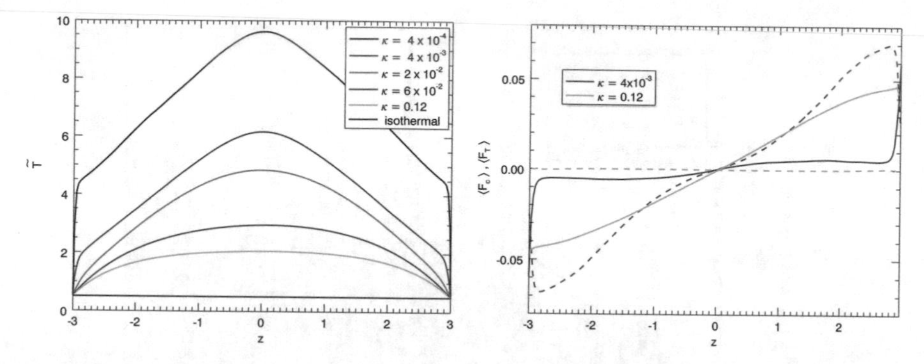

Fig. 1 In the left panel, plot of the temperature averaged on horizontal planes as a function of the vertical coordinate z. The different curves refer to different values of the thermal diffusivity, as shown in the legend. In the right panel, plot of the averaged conductive and convective fluxes as functions of z. The solid curves show the conductive flux, while the dashed curves show the convective flux, the different colors refer to different values of the thermal diffusivity as indicated in the legend

Fig. 2 Temperature distribution at the upper boundary of the computational box

In Fig. 2 we show the temperature distribution on the upper boundary of the computational box, which clearly show a pattern of convective cells. In Fig. 3 we show the time history of Maxwell stresses for cases with different thermal diffusivities, we can see that the transport becomes much more efficient in the convective cases at low diffusivity.

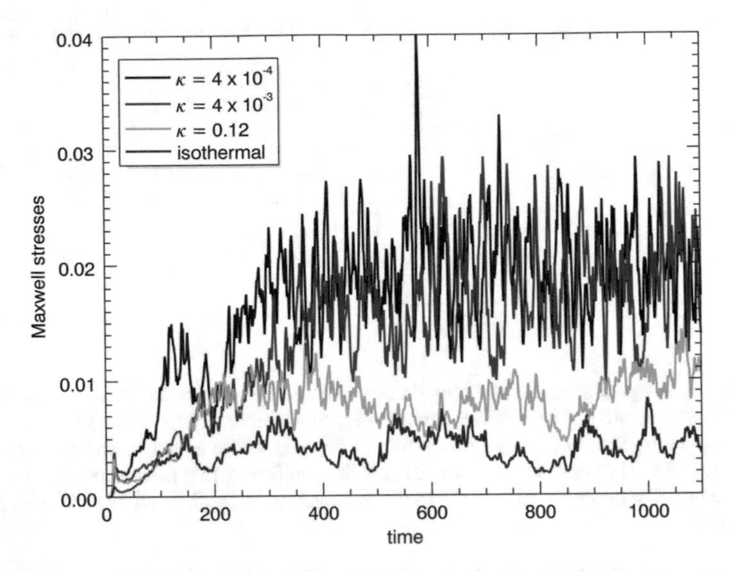

Fig. 3 Time histories of the volume averaged Maxwell stresses for different values of the thermal diffusivity

3 Analytical Models of the Convective Disk Structure

Our aim is to construct the consistent vertical structure of an accretion disk at a given radial position using as a starting point the results of our numerical simulations of shearing boxes with vertical heat transport [10, 11, 13] described in the previous section. There, it is shown that the nondimensional shearing box equations with radiative boundary conditions depend only on two dimensionless parameters, the Péclet number, Pe, which measures thermal diffusivity, and the radiation parameter, Σ_r, which measures the efficiency of the blackbody radiating boundary [13]. The Péclet number and Σ_r are defined as

$$Pe \equiv \frac{D^2 \Omega}{\kappa},$$

$$\Sigma_r \equiv \frac{\sigma \Omega^5 D^5}{\mathcal{R} \rho_0}$$

where D is half the layer thickness, Ω is the rotational frequency, ρ_0 is the average density, σ is the Stefan-Boltzmann constant and $\mathcal{R}$ is the perfect gas constant. In the case of fully convective solutions, at high Pe, the dependence on the Péclet number disappear and we are left with a family of solutions which depends only on Σ_r. For example, the temperature at the top of the convection layer, the heat flux and the turbulent stresses, taking into account the appropriate scaling relations, can be written as

$$T_D = \mathscr{T}(\Sigma_r)\Omega^2 D^2 ,$$

and

$$H = \mathscr{H}(\Sigma_r)\rho_0\Omega^3 D^3$$

$$S = \mathscr{S}(\Sigma_r)\rho_0\Omega^2 D^2$$

The functions $\mathscr{T}(\Sigma_r)$, $\mathscr{H}(\Sigma_r)$ and $\mathscr{S}(\Sigma_r)$ can be determined through numerical investigations, that show that their dependence on Σ_r is very weak [13].

Our procedure starts with building the full vertical structure by taking into account that at the top of the convective region we should have in reality a layer in radiative equilibrium, where the transition between the optically thick and the optically thin regimes occurs. For the radiative layer we can impose several condition: it has to be in hydrostatic equilibrium, its optical depth must be of the order of unity, pressure and density at its base have to match those at the top of the convective region, finally its temperature must be T_{eff}, where

$$\sigma T_{eff}^4 = \mathscr{H}(\Sigma_r)\rho_0\Omega^3 D^3$$

In addition we assume that the thickness of the radiative layer is much smaller than the thickness of the convective regions.

From these relations and imposing the matching condition on pressure, we can obtain the half thickness of the disk D as a function of Ω and ρ_0

$$D = D(\Omega, \rho_0)$$

furthermore, using the equality of the inward angular momentum flux due to accretion to the outward angular momentum flux due to turbulent stresses, we have

$$\rho_0 = \frac{\dot{M}}{2\pi\mathscr{S}D^3\Omega}$$

where $\dot{M}$ is the mass accretion rate. Combining the two relation we can get any property of the disk as functions of the mass accretion rate and Ω.

From the solutions described above, using a power law dependence for the mean opacity, we can check a posteriori their consistency. The solutions were obtained under two assumptions, large Péclet number, $Pe \gg 1$, and scale height of the radiative layer much smaller than D. In Fig. 4 we show in the plane $(\Omega, \dot{M})$, where Ω is given in s^{-1} and $\dot{M}$ is given in g/s, the region where our two assumptions are verified. In the pink region only the high Péclet number condition is verified, in the green region both assumptions are valid. The star correspond to one of the convective solutions found by [12] .

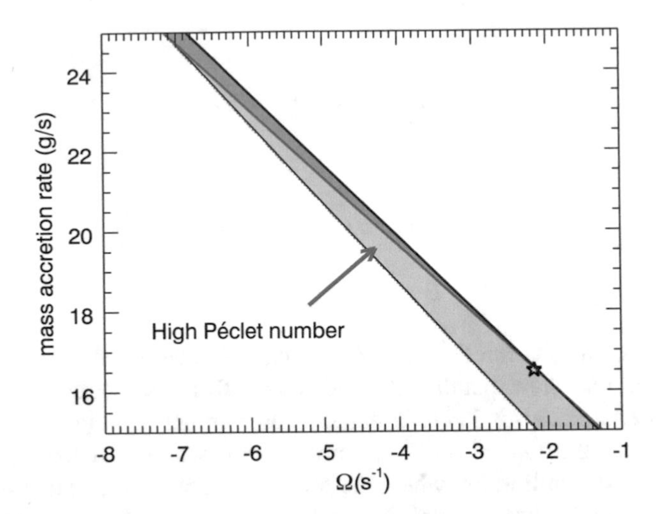

Fig. 4 Regions of model consistency in the $(\Omega, \dot{M})$ plane. The pink region corresponds to convective solutions ($Pe \gg 1$). The green region to convective solutions with thin radiative layers. The star corresponds to one of the convective solutions found by [12]

4 Conclusions

We have performed numerical simulations where we take into account the heat generated by the dissipation of MRI driven turbulence, this is carried by conduction to the surface layers where it is radiated away. If the thermal diffusivity is low, a convective regime may occur, in which heat is carried by convection. In this regime we observe a very efficient large scale dynamo and very effective angular momentum transport. We presented also a procedure for deriving the global disk structure from our local results for fully convective disks. We can then check a posteriori the consistency of our solutions and we can conclude that the fully convective state can extend only to a small portion of the disk, but can be present in any disk. There may be also regions where convection is limited to a central layer, with more extended radiative regions. An interesting point to investigate is what would happen to the more efficient convective regions surrounded by less efficient radiative regions in a global disk.

Acknowledgements We acknowledge that the results in this paper have been achieved using the PRACE Research Infrastructure resource FERMI based in Italy at the Cineca Supercomputing Center.

References

1. S.A. Balbus, J.F. Hawley, ApJ **376**, 214 (1991).
2. S. Fromang, J. Papaloizou, A&A **476**, 1113 (2007).
3. M.E. Pessah, C. Chan, D. Psaltis, ApJ Lett **668**, L51 (2007).
4. P.J. Käpylä, M.J. Korpi, MNRAS **413**, 901 (2011).
5. G. Bodo, F. Cattaneo, A. Ferrari, A. Mignone, P. Rossi, ApJ **739**, 82 (2011).
6. S.W. Davis, J.M. Stone, M.E. Pessah, ApJ **713**, 52 (2010).
7. J. Shi, J.H. Krolik, S. Hirose, ApJ **708**, 1716 (2010).
8. O. Gressel, MNRAS **405**, 41 (2010).
9. X. Guan, C.F. Gammie, ApJ **728**, 130 (2011).
10. G. Bodo, F. Cattaneo, A. Mignone, P. Rossi, ApJ **761**, 116 (2012).
11. G. Bodo, F. Cattaneo, A. Mignone, P. Rossi, ApJ Lett. **771**, L23 (2013).
12. S. Hirose, O. Blaes, J.H. Krolik, M.S.B. Coleman, T. Sano, ApJ **787**, 1 (2014).
13. G. Bodo, F. Cattaneo, A. Mignone, P. Rossi, ApJ **799**, 20 (2015).
14. A. Mignone, G. Bodo, S. Massaglia, T. Matsakos, O. Tesileanu, C. Zanni, A. Ferrari, ApJ Suppl. **170**, 228 (2007).

An Energetic Criterion for Astrophysical Winds and Jets

K. Tsinganos, M. Damoulakis, C. Sauty, and V. Cayatte

1 Collimated Jets and Uncollimated Winds

The first collimated large scale cosmical outflow was observed 100 years ago (1918) by H.D. Curtis in the giant elliptical (E01) galaxy M87, located near the center of the Virgo Cluster. He noted a "curious straight ray" in M87 which was "apparently connected with the nucleus by a thin line of matter" [1]. This observation marked the first discovery of an extragalactic jet. However, the physical nature of this jet would remain obscure for several more decades, awaiting the emergence of radio astronomy in the 1950s, wherein Virgo A would be one of the first discrete sources discovered, and where the "curious" feature was quickly recognized by [2], who were the first to use the term "*jet*". Today the M87 jet is well observed in the radio-to-X-ray frequencies, with indications for an initial jet collimation starting on scales of several Schwarzchild radii for a several billions solar masses black hole, i.e., less than or around 1000 A.U., working outward to scales $\sim$50 kpc [3].

The first Herbig-Haro (HH) object was observed by Burnham around (1890–1894). Specifically, he observed the star T Tauri with the 36-inch (91 cm) refracting

K. Tsinganos (✉)
National and Kapodistrian University of Athens, Section of Astrophysics, Astronomy and Mechanics, Department of Physics, Athens, Greece

National Observatory of Athens, Athens, Greece
e-mail: tsingan@phys.uoa.gr

M. Damoulakis
Department of Physics, National and Kapodistrian University of Athens, Athens, Greece

C. Sauty · V. Cayatte
Laboratoire Univers et Théories, Observatoire de Paris, UMR 8102 du CNRS, Université Paris Diderot, Meudon, France
e-mail: csauty@obspm.fr; Veronique.Cayatte@obspm.fr

C. Sauty (ed.), *JET Simulations, Experiments, and Theory*, Astrophysics and Space Science Proceedings 55,
https://doi.org/10.1007/978-3-030-14128-8_2

telescope at Lick Observatory and noted a small patch of nebulosity nearby; it was catalogued as an emission nebula, later becoming known as Burnham's Nebula (or, HH 255). In the 1940s, A. Joy compiled the first lists of T Tauri stars, i.e., irregular variable stars associated with dark clouds and bright nebulosities, F5-G5 spectral types and low luminosity [4, 5, 7]. Herbig [6] and Haro [8, 9] discovered bright knots of nebulosity in several star-forming regions. This realisation, which is now already 70 years old, initiated the study of star formation in dark molecular clouds. Modern observations demonstrate that HH objects, jets, and outflows are an important and perhaps necessary part of low mass star formation. In fact, the small patches of bright emission which comprise the HH objects have been understood as the result of the interaction of a collimated outflow with the interstellar medium (for reviews see the volume [10]).

Other collimated astrophysical outflows include those from neutron stars/pulsars, such as in the Crab and Vela pulsar and also from galactic X-ray binaries. These contain a compact object, either a neutron star or a stellar-mass black hole, accreting matter from the companion star. In HMXB the companion can be a O or B star, or a blue supergiant [12]; in LMXB the companion is less massive than the compact object and can be on the main sequence, a white dwarf, or a red giant [11]. The accreted matter carries angular momentum and on its way to the compact object usually forms an accretion disk, responsible for the X-ray emission. Relativistic collimated outflows are associated with radio-jet X-ray binaries, the so-called microquasars which often show apparent superluminal motion. Noteworthy microquasars include SS 433 [13], GRS 1915+105 [14] and the very bright Cygnus X-1 [15].

On the other hand, we also have uncollimated cosmical outflows, those termed *winds*, starting with their prototype the solar wind. In this connection, more than a century ago, observations of a continuous auroral activity [16], a periodic 27-day geomagnetic activity [17] and the anti-solar orientation of comet tails [18–20] led Parker to develop the first model of the solar wind as a natural consequence of the basic MHD equations [21]. Subsequently (1958–1964), in-situ observations by spacecrafts verified the existence of the solar wind which fills the Heliosphere in all directions. The wind becomes supersonic at several solar radii, reaching a speed of several hundred km/s with a Mach number $M \simeq 4$ and an angle φ of the Parker spiral of the magnetic field with the radial direction of $\varphi \simeq 45^\circ$ at 1 IAU. The corresponding mass loss of the solar wind outflow is about 10^6 tons per sec, or, $10^{-14}\,M_\odot$/yr.

The task for us now is to understand how, when and why an astrophysical outflow from a gravitating central body becomes collimated, or not, as it propagates at large distances from the central source. This is explored in this paper.

2 A Global Integral in Meridionally Self-Similar HD Outflows

In [22] the basic hydrodynamic equations for the conservation of mass, momentum and energy,

$$\nabla \cdot (\rho \mathbf{V}) = 0\,, \quad \rho(\mathbf{V}\cdot\nabla)\mathbf{V} = -\nabla P - \tfrac{\rho G M}{r^2}\hat{e}_r \tag{1}$$

$$3\left(\tfrac{k\rho}{m_p}\right)(\mathbf{V}\cdot\nabla)T - 2\left(\tfrac{kT}{m_p}\right)(\mathbf{V}\cdot\nabla)\rho = q\,, \tag{2}$$

have been integrated by making the following assumptions. **First**, we express the velocity field via the stream function Ψ,

$$4\pi\rho\mathbf{V} = \nabla \times \left[\frac{\Psi(A)}{r\sin\theta}\hat{e_\phi}\right] + 4\pi\rho V_\phi \hat{e_\phi}\,, \quad A = f(R)\sin^2\theta\,, \tag{3}$$

where we have additionally assumed that the dependence of the stream function $\Psi(R,\theta)$ on R and θ is via the function $A = f(R)\sin^2\theta$, with $R = r/r_o$.
Second, we assume for the free integrals of the mass loss function $d\Psi(A)/dA = 2\pi r_0^2 \rho_0 V_0 \sqrt{1+\delta A}$ and the total angular momentum $L(A) = L_0 A/\sqrt{1+\delta A}$, with δ and L_o constants, and V_0 , ρ_0 the corresponding base values. Then, the flow speed is:

$$V_r(R,\theta) = V_0 \frac{Y(R)f(R)cos\theta}{\sqrt{1+\delta f\sin^2\theta}}\,, \quad V_\phi(R,\theta) = \lambda V_0 \frac{f(R)}{R}\frac{\sin\theta}{\sqrt{1+\delta f\sin^2\theta}}\,, \tag{4}$$

$$V_\theta(R,\theta) = -V_0 \frac{Y(R)R}{2}\frac{df}{dR}\frac{\sin\theta}{\sqrt{1+\delta f\sin^2\theta}}\,. \tag{5}$$

for some function $Y(R)$. Furthermore, in order to separate the variables we find that the required forms for $\rho(R, A)$ and $P(R, A)$ are:

$$\rho(R,A) = \frac{\rho_0}{R^2 Y(R)}[1+\delta A]\,, \quad P(R,A) - P_o = \frac{\rho_o V_o^2}{2} Q(R)[1+\kappa A]\,, \tag{6}$$

where κ is a constant and P_o is the constant pressure at $R \to \infty$. Then, force balance in the meridional plane gives the following set of three equations:

$$\tfrac{dQ}{dR} + \tfrac{2Yf}{R^2}\tfrac{df}{dR} + \tfrac{2f^2}{R^2}\tfrac{dY}{dR} + \tfrac{\nu^2}{YR^4} = 0\,, \tag{7}$$

$$\kappa\tfrac{df}{dR}Q + \kappa\tfrac{dQ}{dR}f - \tfrac{2f^2}{R^2}\tfrac{dY}{dR} - \tfrac{Yf}{R^2}\tfrac{df}{dR} - \tfrac{Y}{2R}\left[\tfrac{df}{dR}\right]^2 + \tfrac{\delta\nu^2 f}{YR^4} - \tfrac{2\lambda^2 f^2}{YR^5} = 0\,, \tag{8}$$

$$\kappa f Q - \tfrac{f}{2}\tfrac{dY}{dR}\tfrac{df}{dR} - \tfrac{fY}{R}\tfrac{df}{dR} - \tfrac{fY}{2}\tfrac{d^2f}{dR^2} + \tfrac{Y}{4}\left[\tfrac{df}{dR}\right]^2 - \tfrac{\lambda^2 f^2}{YR^4} = 0\,. \tag{9}$$

3 Derivation of an Integral of the Equations

We note that in the first Eq. (7) appears the derivative of the dimensionless pressure function, $dQ(R)/dR$, in the third Eq. (9) appears the dimensionless pressure function Q(R) itself, and in the second Eq. (8) we have both Q(R) and $dQ(R)/dR$. We shall try to extract a single equation from the above system by eliminating the function Q(R). This can be easily done by solving the first equation for $\kappa f dQ/dR$, the third equation for $\kappa (df/dR)Q$, and replace the results into the second equation:

$$\frac{1}{2}\frac{dY}{dR}\left[\frac{df}{dR}\right]^2 + \frac{Y}{R}\left[\frac{df}{dR}\right]^2 + \frac{Y}{2}\frac{df}{dR}\frac{d^2f}{dR^2} - \frac{Y}{4f}\left[\frac{df}{dR}\right]^3 + \frac{\lambda^2 f}{YR^4}\frac{df}{dR} - \frac{2\kappa Y f^2}{R^2}\frac{df}{dR} \tag{10}$$

$$-\frac{2\kappa f^3}{R^2}\frac{dY}{dR} - \frac{\kappa \nu^2 f}{YR^4} - \frac{2f^2}{R^2}\frac{dY}{dR} - \frac{Yf}{R^2}\frac{df}{dR} - \frac{Y}{2R}\left[\frac{df}{dR}\right]^2 + \frac{\delta\nu^2 f}{YR^4} - \frac{2\lambda^2 f^2}{YR^5} = 0\,. \tag{11}$$

Multiply the above equation with YR^2/f and rearrange the various terms to get:

$$\underbrace{\frac{R^2}{2f}Y\frac{dY}{dR}\left[\frac{df}{dR}\right]^2 + \frac{Y^2R}{2f}\left[\frac{df}{dR}\right]^2 + \frac{Y^2R^2}{2f}\frac{df}{dR}\frac{d^2f}{dR^2} - \frac{R^2Y^2}{4f^2}\left[\frac{df}{dR}\right]^3}_{\frac{d}{dR}\left[\frac{Y^2R^2}{4f}\left[\frac{df}{dR}\right]^2\right]} - \tag{12}$$

$$\underbrace{2\kappa Y^2 f\frac{df}{dR} - 2\kappa f^2 Y\frac{dY}{dR}}_{\frac{d}{dR}(-\kappa f^2Y^2)} + \underbrace{\frac{\lambda^2}{R^2}\frac{df}{dR} - \frac{2\lambda^2 f}{R^3}}_{\frac{d}{dR}\left[\frac{\lambda^2 f}{R^2}\right]} \underbrace{-\frac{\kappa\nu^2}{R^2} + \frac{\delta\nu^2}{R^2}}_{\frac{d}{dR}\left[\frac{(\kappa-\delta)\nu^2}{R}\right]} \underbrace{-Y^2\frac{df}{dR} - 2fY\frac{dY}{dR}}_{\frac{d}{dR}(-fY^2)} = 0\,. \tag{13}$$

Evidently, the above equation can be expressed as the derivative of a single quantity:

$$\frac{d}{dR}\left[\frac{Y^2R^2}{4f}\left(\frac{df}{dR}\right)^2 - \kappa f^2Y^2 - fY^2 + \frac{\lambda^2 f}{R^2} + \frac{(\kappa-\delta)\nu^2}{R}\right] = 0\,. \tag{14}$$

By integrating this equation we get that the bracketed quantity, let us call it ε, is independent of R, even if its individual terms are functions of R:

$$\varepsilon = \frac{Y^2R^2}{4f}\left[\frac{df}{dR}\right]^2 - \kappa f^2Y^2 - fY^2 + \frac{\lambda^2 f}{R^2} + \frac{(\kappa-\delta)\nu^2}{R} \Rightarrow \frac{d\varepsilon}{dR} = 0\,. \tag{15}$$

An equivalent form of ε can be found in terms of the function $F(R) = Rf(df/dR)$,

$$\varepsilon = Y^2 f\left[\frac{F^2}{4} - 1\right] - \kappa(Yf)^2 + \frac{\lambda^2 f}{R^2} + \frac{(\kappa-\delta)\nu^2}{R}\,. \tag{16}$$

4 An Alternative Definition of the Constant ε as a Global Integral

It is to be noted that independence of the constant ε on R also implies and its independence on any individual streamline A=const. That means, the quantity ε is not only constant on each individual stream line A=const., but it has the same value globally, along any distance R and along all streamlines A=const. This result is a rather important one for our model, considering that every physical quantity of the problem depends on the streamline we examine, while ε is a global, characteristic of the problem. This constant ε is also important because it may be related to the asymptotical shape of the outflow, i.e., if it is a cylindrical jet $F = 2$, or, a conical wind $F = 0$ depending, for example, on whether ε is positive or negative, respectively as in [23].

At this point it is worth to recall that while all known radially self-similar models accept a polytropic relationship between pressure and density (and as a result of that we obtain the familiar Bernoulli integral), meridionally self-similar models do not accept a polytropic relationship between pressure and density, because of the existence of an external heating, so a Bernoulli integral does not exist. It seems therefore that ε is a more general integral that connects the internal energy with the heating reservoir, and that it can replace the somehow artificial (albeit convenient) polytropic relationship between pressure and density,

$$q = \rho \mathbf{V} \cdot \left[\nabla h - \frac{\nabla P}{\rho} \right] = \rho \mathbf{V} \cdot \nabla E\,, \tag{17}$$

where q is the required volumetric heating which is needed in order that the gas is able to escape from the gravitational well of the central object and which is independent of any particular relationship between pressure and density, while E is the sum of the kinetic, thermal (enthalpy), gravitational and Poynting energy flux densities per unit of mass flux density. The resulting heat content, or enthalpy h, represents the internal reservoir from which the gas absorbs the required energy and converts it to kinetic energy at large distances. The enthalpy function h for a monatomic gas with a ratio of the specific heats $\Gamma = c_P/c_V$ is $\Gamma/(\Gamma - 1)kT/\bar{m}$. However, the pressure need not be constrained by an artificial polytropic law, $P \neq K\rho^\gamma$. The only imposed constrain on the pressure function here is that in Eq. (6) relating the ratio of the $\theta-$ dependent part of the gas presssure to its spherically symmetric part, to be κA. Note that if the effective polytropic index γ is less than the ratio of the specific heats Γ, that means heat is provided to the gas by some external source, i.e., external heat and internal work by the gas 'flow' in opposite directions: the gas expanding does work while the surroundings, (or the external source) add energy to the gas.

This claim can be bolstered by the fact that an additional proof for the existence of ε under the previous conditions exists, which is related to the gradient of the modified energy of the system, E_o:

$$\rho \mathbf{V} \cdot \nabla E_o = -\frac{\rho \mathbf{V} \cdot \nabla P}{\rho}, \quad E_o = E - h. \tag{18}$$

The above equation gives,

$$\left.\frac{\partial E_o(R, A)}{\partial R}\right|_{A=const.} = -\frac{1}{\rho(R, A)} \left.\frac{\partial P(R, A)}{\partial R}\right|_{A=const.}. \tag{19}$$

With the density $\rho(R, A)$ and pressure $P(R, A)$ given by Eqs. (6) in the present meridionally self-similar model, wherein additionally κ = const., by substituting this expression in the previous Eq. (19) we get,

$$\left.\frac{\partial E_o(R, A)}{\partial R}\right|_{A=const.} = -\frac{V_o^2}{2} \frac{1+\kappa A}{1+\delta A} Y R^2 \frac{dQ}{dR}, \tag{20}$$

By integrating Eq. (20) we obtain the following expression for $E_o(R, A)$,

$$E_o(R, A) = -\frac{V_o^2}{2} \frac{1+\kappa A}{1+\delta A} \int_{R_o}^{R} Y R^2 \frac{dQ}{dR} dR + E_o(R_o, A), \tag{21}$$

wherein $E_o(R_o, A)$ is a suitable arbitrary function of A coming from the integration in R of Eq. (20). Then, it follows,

$$\left.\frac{\partial E_o(R, A)}{\partial A}\right|_{A=0} + \frac{(\kappa - \delta) V_o^2}{2} \int_{R_o}^{R} Y R^2 \frac{dQ}{dR} dR = \left.\frac{\partial E_o(R_o, A)}{\partial A}\right|_{A=0}. \tag{22}$$

Substituting in the above equation the expression of the integral in R from the previous Eq. (21) applied to $A = 0$ we get,

$$\left.\frac{\partial E_o(R, A)}{\partial A}\right|_{A=0} - (\kappa - \delta) E_o(R, 0) = -(\kappa - \delta) E_o(R_o, 0) + \left.\frac{\partial E_o(R_o, A)}{\partial A}\right|_{A=0}. \tag{23}$$

Since the RHS is a constant, we may write the above Eq. (23) as follows,

$$\left.\frac{\partial E_o(R, A)}{\partial A}\right|_{A=0} - (\kappa - \delta) E_o(R, 0) = \frac{V_o^2}{2} \epsilon, \tag{24}$$

where ε is a dimensionless constant. We may rewrite the above equation by defining a new quantity $\varepsilon\prime$ as follows,

$$\varepsilon\prime = \frac{2}{V_o^2}\frac{\partial E_o(R,A)}{\partial A}\bigg|_{A=0} = \varepsilon + \frac{2}{V_o^2}(\kappa - \delta)E_o(R,0) \equiv \varepsilon + \mu\,, \tag{25}$$

with $\mu \equiv 2(\kappa - \delta)E_o(R,0)/V_o^2$, in an analogous way this quantity has been introduced in [24–27]. Finally, another way to write Eq. (24) is:

$$E_o(R,0)\frac{\partial}{\partial A}\left[\ln\left(\frac{\rho E_o}{P - P_0}\right)\right]\bigg|_{(R,\ A=0)} = \frac{V_o^2}{2}\epsilon\,, \tag{26}$$

or,

$$E_o(R,0)\left[\frac{\partial}{\partial A}\ln E_o - \frac{\partial}{\partial A}\ln(T - T_o)\right]\bigg|_{(R,\ A=0)} = \frac{V_o^2}{2}\epsilon\,. \tag{27}$$

In the previous Eqs. (24–27), we have obtained from first principles a global integral of the problem ε which depends solely on R and which can be calculated a priori from the conditions at the base of the outflow, without a need to know the total input of heating along each line. Note that the sign of ϵ in Eq. (27) comes from the competition of two terms. The first is the variation across the poloidal streamlines/fieldlines of the available energy E_o. When this energy increases as move away from the polar line $A = 0$, this term is positive. The corresponding force, which has the opposite sign to the variation of the energy, $\partial E_0(A)/\partial A$, points towards the polar axis and therefore it produces collimation of the outflow according to Le Chatelier's principle: when any system at equilibrium for a long period of time is subjected to change in concentration, temperature, volume, or pressure, then the system readjusts itself to partly counteract the effect of the applied change and a new equilibrium is established. Thus $\epsilon\prime > 0$ means collimated outflow.

The second competing term in Eq. (27) gives the variation of the enthalpy across the streamlines, as we move away from the polar line $A = 0$. In Eq. (24) this depends on the sign of the constant $(\kappa - \delta)$. In thermally driven winds, this heat content is pivotal for lifting the gas out of the gravitational well of the central object. With the density increasing as we move away from the central axis for positive δ, an increased heating is needed in order to balance the gravitational force, as we move away from the polar line $A = 0$. The remaining heat content is available to drive the wind.

5 Jet or Wind?

We can see now that the equation of the integral ε in the form of Eq. (16) can be simplified to the following forms, for $F = 2$ (jet), or $F = 0$ (wind), respectively:

$$\varepsilon_{jet} = c_1\lambda^2 - \kappa c_1^2 (Y R^2)^2_\infty = \varepsilon\prime_{jet} - \mu = c_1\lambda^2 - \kappa \left(\frac{V_r(R\to\infty,\theta=0)}{V_0}\right)^2 \quad (28)$$

$$\varepsilon_{wind} = -c_2 Y^2_\infty - \kappa (c_2 Y_\infty)^2 = \varepsilon\prime_{wind} - \mu = -c_2 Y^2_\infty - \kappa c_2^2 Y^2_\infty\,, \quad (29)$$

where $c_1 > 0$ is the positive constant in $f(R) = c_1 R^2$ for a jet and $c_2 > 0$ the value of $f(R)$ at infinity for a wind type outflow. Since these quantities ε_{jet} and ε_{wind} are constants everywhere, they will remain constants also for $R \to \infty$, so we have taken their corresponding expressions for $R \to \infty$, omitting terms of the form $const/R^n$. Note that for $\kappa = 0$ we always have $\varepsilon^{jet} > 0$, while $\varepsilon^{wind} < 0$. Then, the *necessary* and *sufficient* condition to have a jet is $\varepsilon > 0$ and a wind $\varepsilon < 0$.

The integral ε can be calculated at the stellar base $R_0 = 1$ wherein $Y_0 = 1,\ f_0 = 1,\ F_1 = s$, in terms of the given system parameters s, λ, κ, δ and ν:

$$\varepsilon = \frac{s^2}{4} - 1 + \lambda^2 - \kappa + (\kappa - \delta)\nu^2\,. \quad (30)$$

Thus, we may predict the asymptotic behaviour of the outflow by knowing the values of the 5 parameters κ, δ, λ, ν and s.

Acknowledgements The authors are indebted to Drs Joao Lima and Nektarios Vlahakis, as well as to Loic Chantry for several stimulating discussions on the subject of this paper. KT would like to express his very great appreciation to all participants of the JETSET network for a pleasant and fruitful collaboration throughout the last 15 years.

References

1. Curtis, H. D.: Publications of Lick Observatory, **13**, pp. 9–42 (1918)
2. Baade, W. and Minkowski, R.: ApJ, **119**, p. 215 (1954)
3. Kim, J.-Y., Krichbaum, T. P., Lu, R.-S., Ros, E., Bach, U., Bremer, M., de Vicente, P., Lindqvist. M. and Zensus, J.A., A&A, (2018) (in press, astro-ph 1805.02478)
4. Joy, A.H., ApJ, **102**, p. 168 (1945)
5. Joy, A.H., ApJ, **110**, p. 424 (1949)
6. Herbig, G.H., ApJ, **113**, p. 697–699 (1951)
7. Herbig, G.H., Adv. Astr. Astrophys., **1**, p. 47–103 (1962)
8. Haro, G., ApJ, **115**, p. 572 (1952)
9. Haro, G., ApJ, **117**, p. 73 (1953)
10. Tsinganos, K., Ray, T., Stute, M., *Protostellar Jets in Context*, Astrophysics and Space Science Proceedings Series. Springer-Verlag Berlin Heidelberg (2009).
11. Remillard R. A., McClintock J. E., 2006, ARA&A, *44*, p. 49 (2006).
12. Sidoli, L., Paizis, A., MNRAS, (2018) (in press, astro-ph 1809.00814)

13. Bodo, G., Ferrari, A., Massaglia, S.,Tsinganos, K., A&A, *149*, p. 246–252 (1985)
14. Tetarenko, A. J., Freeman, P., Rosolowsky, E. W., Miller-Jones, J. C. A., Sivakoff, G. R., MNRAS (2018) (in press, astro-ph 1712.00432)
15. Axelsson, M., Done, C., MNRAS, (2018) (in press, astro-ph:1803.01991)
16. Birkeland, K., The Norwegian Aurora Polaris Expedition 1902–1903, Norway, Christiania: Aschehoug, *1*, pp. 1–801, (1908, 1913).
17. Chapman, S., Bartels, J.: Geomagnetism, Vol. I: Geomagnetic and Related Phenomena, London: Oxford Univ. Press (1940)
18. Biermann, L., Zeitschrift fur Astrophysik, *29*, p. 274 (1951)
19. Biermann, L., Zs.f. Naturforsch., *7a*, p. 127 (1952)
20. Biermann, L.,1957, Observatory, *107*, 109 (1957)
21. Parker, E. N., ApJ, *128*, p. 664 (1958)
22. Tsinganos, K., Sauty, C., A&A, *255*, p. 405–419, (1992)
23. Sauty, C., Tsinganos, K., A&A, *287*, p. 893–926, (1994)
24. Sauty, C., Tsinganos, K., Trussoni, E., A&A, *348*, p. 327–349 (1999)
25. Sauty, C., Trussoni, E., Tsinganos, K., A&A, *389*, p. 1068–1085 (2002)
26. Sauty, C., Trussoni, E., Tsinganos, K., A&A, *421*, p. 797–809 (2004)
27. Sauty, C., Trussoni, E., Tsinganos, K., A&A, *533*, id.A46, 12 pp. A46 (2011)

Meridional Self-Similar MHD Relativistic Flows Around Kerr Black Holes

L. Chantry

Recent observational projects, e.g., ALMA and γ-ray telescopes such as HESS and HESS2 (also in the future the CTA) have provided new observational constraints on the central region of rotating black holes in AGN, suggesting that there is an inner- or spine-jet surrounded by a disk wind. The EHT can also image the vicinity of the black hole at radio and millimeter wavelengths. The disk-wind component, which is probably composed of a powerful hadronic plasma, most likely originates from the accretion disk via the magneto-centrifugal launching mechanism. High resolution radio imaging of AGN have revealed that some sources present motion of superluminal knots starting very close to the central object. They also shows that outflows originate in regions where General Relativity (GR) cannot be neglected. Observations also indicate a transverse stratification of their jets.

The spine-jet component is likely to be composed of electron – positron pairs (leptons), which are formed in the environment of the black hole. The pair creation is induced by very energetic particles (neutrinos – photons) coming from the disk. The electro-magnetics fields of the spine jet field-line can extract angular momentum and energy from the black hole. Then the power of the spine jet could contains the energy flux extracted from the black hole and the energy flux injected during the pairs formation.

The first motivation of this work is to get a spine jet model able to describe and propose a dynamical explanation (launching/collimation) of these outflow. The second is to use this model to build solution till the black hole horizon field (inflow and outflow from loaded material), in order to explore the power origin of these flow. Injected power via paris formation or extracted power from the black hole?

L. Chantry (✉)
Laboratoire Univers et Théories, Observatoire de Paris-PSL, UMR 8102 du CNRS, Université Paris Diderot, Meudon, France
e-mail: loic.chantry@obspm.fr

C. Sauty (ed.), *JET Simulations, Experiments, and Theory*, Astrophysics and Space Science Proceedings 55,
https://doi.org/10.1007/978-3-030-14128-8_3

1 Meridional Self-Similar Models for MHD Flow Around Kerr Black Hole

The role of magnetic fields in such flows is today well admitted. The question is to describe a phenomenon, which mixes gravity and electro-magnetism. To describe the spine jet, we choose a fluid approach, which leads to study the magneto-hydrodynamics (MHD) field in the environment of a supermassive black hole. To avoid useless complication but keep possibilities to have some energetic exchange with the black hole via the Poynting flux (or generalized Penrose mechanism), we chose the Kerr geometry to fix the gravitational field. Indeed, frame dragging effects can become non-negligible in the close environment of the black-hole. To obtain the MHD equations, we used the 3+1 formalism to decompose all physical quantities in the frame of a specific observer, the zero angular momentum observer (ZAMO). We derive the fundamental equations (Continuity, Maxwell, Euler equations and the first principle of thermodynamics). After that, we used the usual main result of axisymmetric, stationary, ideal MHD in Kerr metrics (First integrals, Alfvén transition critical surface…).

To build our model, we assume that the poloidal Mach number and pressure are separable functions of radius and magnetic flux. Conversely to the classical jet model of [4], where they have self similar exact solutions, we make here an expansion of all quantities to second order in colatitude. Our magnetic flux function has meridional self-similarity. In order to expand the poloidal Euler equation, we use the constraints on the critical Alfvén surface, supposed to be spherical, but not neglecting the effect of the light cylinder conversely to all previous meridional self-similar models of relativistic spine jets. This introduces a set of seven dimensionless parameters and three unknown functions of the radius. These parameters and these functions are able to describe all physical quantities at second order in the expansion. Our system of equation possess a second critical surface, the so-called slow-magneto-sonic. We use a Runge-Kutta solver. I built an automatic numerical research, that for a given set of parameters, cross all critical surfaces.

As in previous self-similar flows model [4] in Newtonian and [3] in Schwarzschild gravity, we built a generalization of the so-called magnetic collimation efficiency, ϵ. This is a constant parameter of the model that quantifies how much the magnetic field can collimate the flow. The model is able to produce two kinds of flows, outflows with positive outward mass flux and inflows with negative mass flux. The Inflow (outflow) describe the field before (up to) a radius, that we called stagnation radius, where the poloidal speed of the flow becomes zero.

2 Results

2.1 *Outflow*

This model of flow include, on the second order in colatitude, the effect of light cylinder radius, which as been neglect in previous models as previously mentioned. For outflow solution, these additional terms practically act as an extra transversal pressure. This excess of collimation force also induces strongest initial acceleration.

We can find solutions, i.e. Fig. 1, with really high γ Lorentz factors ($\approx$10–100). These kind of solutions is described in [1]. The acceleration is mainly due to a transfer of energy from a specific effective enthalpy to kinetic energy, these solutions are mainly enthalpy driven. In the solution (K2) the collimation is mainly due to toroidal magnetic force, where the transversal balance is dominated by electric and magnetic forces. The total electro-magnetic transversal forces have the same order as pressure transversal forces.

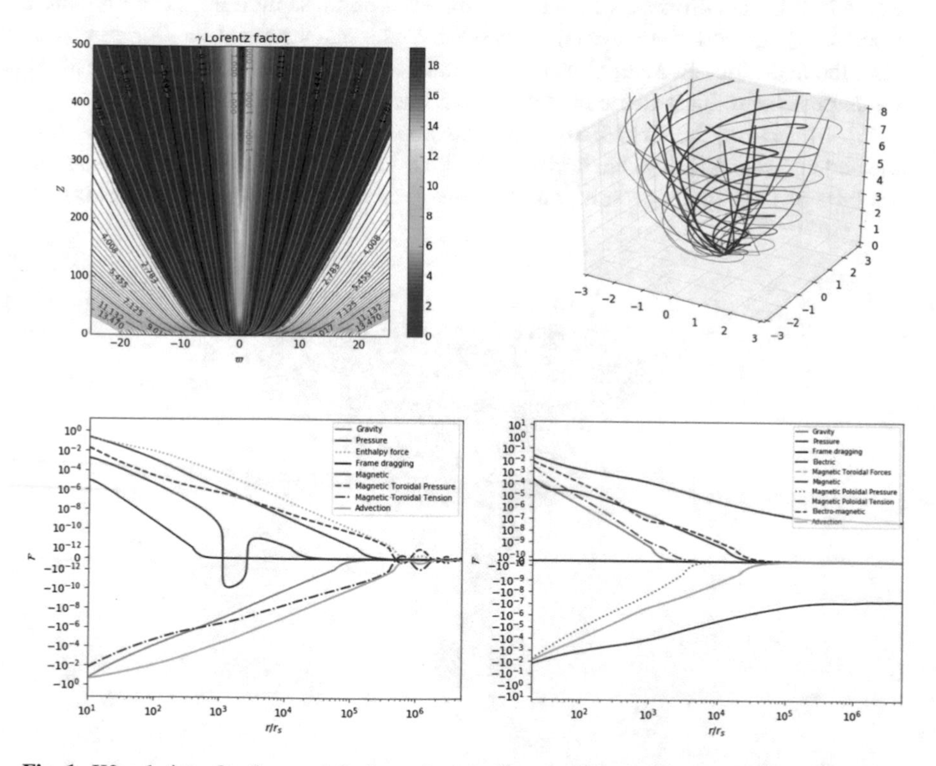

Fig. 1 K2 solution. On the top left, the γ Lorentz factor of the flow in the poloidal plane. The poloidal field line are represented in black. On the top right, 3 dimensional representation of field line. In red the magnetic field line and in blue the velocity field line. On the bottom left, the longitudinal force along a poloidal fieldline line in the flow. And in the bottom right, the transversal force along a poloidal field line in the flow

At infinity, the outflow can be conical or cylindrical. Cylindrical outflow are adapted to models AGN. In function of the parameter, for cylindrical outflow, the pressure can give a significant contribution to the collimation (K1) but for the most part of them, they are magnetically collimated (K2–K3). An interesting solution we get, is the solution K3, which is a quite good models of M87 spine jet if we combine this solution with disk wind component. The conical outflow solutions could be adapted to models outflows of Seyfert Galaxies or GRB. Accepting the relation between magnetic flux and accretion rate [5], we are able to calculate the efficiency of the jets of our model. For the solution K2 and K3 we obtain efficiency around ≈40–50%.

2.2 *Double Flow*

The only way for field line to connect the black hole horizon and the outflow region is to have a process, which injects material on these field lines (Pair creation process....). Indeed, the material must fall into the black hole at the horizon. And since the mass flux on a field line must remain constant without material injection, it is impossible to have an ejection on such a line without material injection.

With injection, the field lines can have a double flow structure, some of this injected material falls into the horizon and the other part is ejected, Fig. 2. To model this type of field line, one solution consists, as in the work of [2], in matching an

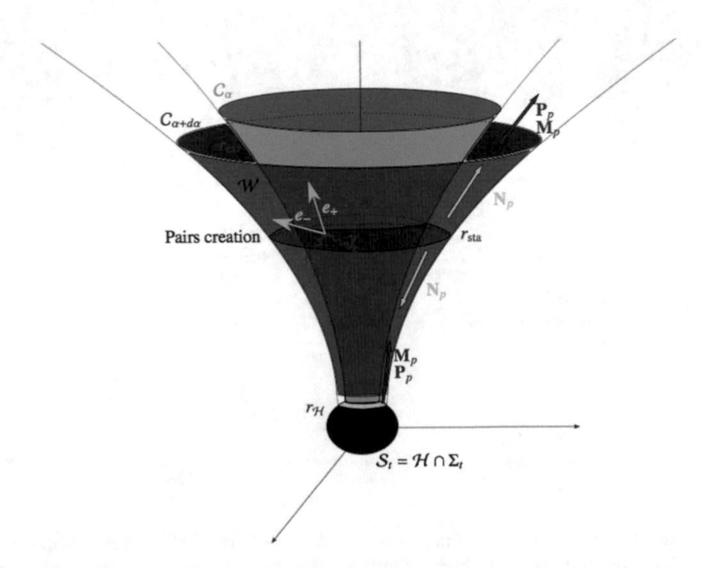

Fig. 2 Schematic representation of the double flow structure tubes used in this work. The pairs are injected on the stagnation surface (red). Then some of the injected material exits the system and the rest falls to the horizon, $\mathbf{N}_p$ is the current number, $\mathbf{P}_p$ the Noether energy flux and $\mathbf{M}_p$ the angular momentum energy flux can exit the system in certain MHD field configuration

inflow solution and an outflow solution with Dirac load terms (perhaps pairs creation or other mechanism) injecting material at the common stagnation surface. Matter come from the loading terms region at the stagnation surface to flow in and out of the system, then we obtain an entire description of the field from horizon to large distances.

In this configuration, we could suppose that the energy flux on the black hole is only composed of the plasma part and electromagnetic part

The energy flux exchange with the black hole could be decompose in three terms,

$$\Phi_{\mathrm{M}}\big|_{\mathscr{H}} = -\frac{M^2_{\mathrm{Alf}}\big|_{\mathscr{H}}}{M^2_{\mathrm{Alf}}\big|_{\mathscr{H}} + \varpi^2_{\mathscr{H}}(\Omega - \omega_{\mathscr{H}})^2/c^2}\left(\Psi_{\mathrm{A}} L \omega_{\mathscr{H}} - \Psi_{\mathrm{A}}\mathscr{E}\right)$$

$$\Phi_{\mathrm{LT}}\big|_{\mathscr{H}} = \Psi_{\mathrm{A}}\mathscr{L}\omega_h + \frac{\varpi^2_{\mathscr{H}}\omega_{\mathscr{H}}(\Omega - \omega_{\mathscr{H}})/c^2}{M^2_{\mathrm{Alf}}\big|_{\mathscr{H}} + \varpi^2_{\mathscr{H}}(\Omega - \omega_{\mathscr{H}})^2/c^2}\left(\Psi_{\mathrm{A}} L \omega_{\mathscr{H}} - \Psi_{\mathrm{A}}\mathscr{E}\right)$$

$$\Phi_{\mathrm{EM}}\big|_{\mathscr{H}} = \frac{\varpi^2_{\mathscr{H}}\Omega(\omega_{\mathscr{H}} - \Omega)/c^2}{M^2_{\mathrm{Alf}}\big|_{\mathscr{H}} + \varpi^2_{\mathscr{H}}(\Omega - \omega_{\mathscr{H}})^2/c^2}\left(\Psi_{\mathrm{A}} L \omega_{\mathscr{H}} - \Psi_{\mathrm{A}}\mathscr{E}\right)$$

the inertial part Φ_M contains enthalpy, kinetic and gravitational energy flux per unit of magnetic flux. Lense-Thirring part Φ_{LT} is link to frame-dragging and the electromagnetic part Φ_{EM} simply is the Poynting flux par unit of magnetic flux. The null energy condition insure that Φ_{M} remain negative on the black hole horizon, then the two other flux could be positive in some configuration of fields. The optimal value of Ω which optimize Poynting flux on the black hole is $\omega_{\mathscr{H}}/2$. We also can prove that to extract black hole rotational energy in the hydrodynamical case, the stagnation radius of inflow need to be inside of the ergosphere. The calculation of inflow solutions of the model indicate three kind of solution Fig. 3. One kind of solution (I1) is dominated everywhere by the "mass energy" flux and do not extract rotational energy. The second kind (I2) is dominated somewhere on the horizon by the "Lense-Thirring" energy flux (Penrose mechanism). The last kind

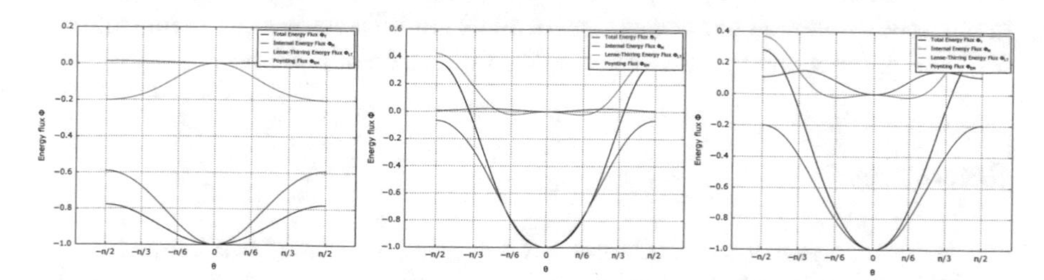

Fig. 3 Plot of the inertial Φ_M, Lense-Thirring Φ_{LT} and electro-magnetic Φ_{EM} and the total flux Φ_T normalized energy fluxes by unit of magnetic flux on the black hole horizon as a function of the colatitude angle θ for solution I1 (left), I2 (middle) and I3 (right). All the flux are normalized by the total flux on the polar axis $\Phi_T(0)$

of solution (I3) are dominated somewhere by the Poynting flux (Blandford-Znajek mechanism). Nevertheless the field line geometry leads us to think these inflow solution could be valid on the black hole horizon for latitude less than $\pi/3$.

We calculated three outflow solutions, which respect minimal matching conditions. Thus, we are able to calculate ad hoc load terms allowing us to obtain a physically consistent solution. For the moment, the power source of outflow of our solution we obtain are dominated by the loading terms. The angular momentum is extracted from the black hole in each of these solutions. We also derive from the model, a scaling law comparable to those explore in [5] except it link the magnetic flux threading the black and the pair mass infall rate and not the total accretion.

3 Conclusions

We built a meridional self similar model around Kerr black holes allowing us to calculate solutions of GRMHD from the black hole horizon to infinity. This allow us to explore dynamics and energetics issue of these flow.

We plan to test the stability of my model solutions on some simulation, using `PLUTO` and `AMR-VAC` code. First of all we need to have an estimation of steady state and dynamical time scale, in order to determine the region of the flow that could be tested in simulation. And then to complete our spine jet components with disk-wind one to study their interaction.

The necessary loading terms used to match our solution needs to be compare with realistic physical process value such as pairs creation rate or others process. The created pairs originate from neutral particles (photon or neutrino) originate from the disk. Then the loading terms are linked to the disk physical conditions. This aspect need also to be explore in order to estimate these rate.

Because of the separation of variables, we obtain non-polytropic flows, which are sustained by a generalized pressure requiring additional heating. In other words the dynamics controls the thermodynamics. The nature of this extra energy needs to be explained. It can be interpreted in terms of radiative or turbulent MHD waves in the flow, or with an out of thermodynamical equilibrium gas (but with isotropic distribution function of particle speed in the fluid referential frame).

References

1. Chantry, L.,Cayatte, V.,Sauty, C.,Vlahakis, N.,Tsinganos, K.. A&A, 612:A63, April 2018
2. Globus, N.,Levinson, A.. PRd, 88(8):084046, October 2013.
3. Meliani, Z.,Sauty, C.,Vlahakis, N.,Tsinganos, K.,Trussoni, E.. A&A, 447:797–812, March 2006.
4. Sauty, C.,Tsinganos, K., A&A, 287:893–926, July 1994.
5. M. Zamaninasab, E. Clausen-Brown, T. Savolainen, and A. Tchekhovskoy. Nature, 510:126–128, June 2014

Part II
Simulations

Modelling the Accretion on Young Stars, Recent Results and Perspectives

L. de Sá, C. Stehlé, J. P. Chièze, I. Hubeny, T. Lanz, S. Colombo, L. Ibgui, and S. Orlando

1 Accretion in Context

Classical T Tauri Stars (CTTSs) are low mass pre-main sequence stars accreting material from the surrounding thick disk. Following the magnetospheric scenario, the infalling plasma is funneled at the free-fall speed (~100 km/s) through the stellar magnetosphere by an intense field (>1kG, see e.g. [10]), and is stopped in the stellar atmosphere where the flow ram pressure equals the local thermal pressure: a

L. de Sá (✉) · J. P. Chièze
LERMA, Sorbonne Université, Observatoire de Paris, Université PSL, CNRS, Paris, France

CEA/DSM/IRFU/SAp-AIM, CEA Saclay, CNRS, Gif-sur-Yvette, France
e-mail: lionel.desa@obspm.fr

C. Stehlé · L. Ibgui
LERMA, Sorbonne Université, Observatoire de Paris, Université PSL, CNRS, Paris, France
e-mail: chantal.stehle@obspm.fr; laurent.ibgui@obspm.fr

I. Hubeny
Steward Observatory, University of Arizona, Tucson, AZ, USA

T. Lanz
Observatoire de la Côte d'Azur, Nice, France
e-mail: thierry.lanz@oca.eu

S. Colombo
INAF-Osservatorio Astronomico di Palermo, Palermo, Italy

LERMA, Sorbonne Université, Observatoire de Paris, Université PSL, CNRS, Paris, France
e-mail: salvatore.colombo@obspm.fr

S. Orlando
INAF-Osservatorio Astronomico di Palermo, Palermo, Italy
e-mail: orlando@astropa.inaf.it

C. Sauty (ed.), *JET Simulations, Experiments, and Theory*,
Astrophysics and Space Science Proceedings 55,
https://doi.org/10.1007/978-3-030-14128-8_4

strong shock forms and a heated plasma slab develops upwind. The observed X-ray emission indicates the presence of dense ($n_e \sim 10^{11}\,\mathrm{cm}^{-3}$) and hot ($T > 1$ MK) plasma characterizing the post-shock medium [11].

1D simulations predict quasi-periodic oscillations (QPOs) of the post-shock structure with typical period of several hundreds of seconds [24, 25] for an accretion flow velocity of $400\,\mathrm{km/s}$ impacting a static atmosphere. In these works, the medium is assumed to be optically thin: the absence of radiative absorption prevents any radiation feedback between the hot accretion column and the surrounding media. However, the predicted QPOs remain undetected [7, 8].

The topology of the accretion column [19–21], its dynamics versus the magnetic field and inflow velocity [14, 16, 17] as also the expected X-ray signatures [1, 5, 6, 25] have been numerically explored through 2D MHD simulations to reconcile models with observations (the discrepancy between X-ray and Vis/UV derived accretion rates for instance). We consider in this work the problem of the interplay between the chromosphere and the accretion stream on one side and to inspect more carefully the role played by radiation using 1D radiation hydrodynamics simulations.

2 1D Radiation Hydrodynamics

In the strong magnetic field case, the post-shock plasma is not building uniformly but is expected to be fragmented into independent fibrils that oscillate out of phase [14, 15] into which the magnetic field plays no role. In our study, we have performed 1D simulations of such a fibril and the stellar atmosphere beneath it along one magnetic field line, neglecting its effect on the structure. However, we have taken care of the coupling between radiation and matter. We present in this section the physical model and the numerical tools we used to this purpose. The numerical results will be presented in the next section.

The plasma (mass density ρ, temperature T, internal energy density e, velocity $\mathbf{v}$, pressure p) is described by the hydrodynamics equations and the two momenta (the radiation energy and flux/momentum densities, resp. E_r and $\mathbf{F}_r = c^2\,\mathbf{M}_r$) equations for radiation transfer (written in the comoving frame) [13]:

$$\begin{cases} \partial_t \rho + \nabla \cdot (\rho\,\mathbf{v}) = 0 \\ \partial_t \rho\mathbf{v} + \nabla \cdot (\rho\,\mathbf{v} \otimes \mathbf{v}) + \nabla p = \mathbf{s}_m \\ \partial_t e + \nabla \cdot (e\,\mathbf{v}) + p\,\nabla \cdot \mathbf{v} = s_e \\ \partial_t E_r + \nabla \cdot (E_r\,\mathbf{v}) + (\mathsf{P}_r : \nabla) \cdot \mathbf{v} + \mathbf{v} \cdot \partial_t \mathbf{F}_r / c^2 + \nabla \cdot \mathbf{F}_r = s_{E_r} \\ \partial_t \mathbf{M}_r + \nabla\,(\mathbf{M}_r \cdot \mathbf{v}) + (\mathbf{M}_r \cdot \nabla)\,\mathbf{v} + \mathbf{v} \cdot \partial_t \mathsf{P}_r / c^2 + \nabla \cdot \mathsf{P}_r = \mathbf{s}_{\mathbf{M}_r} \end{cases} \tag{1}$$

along with the pseudo-perfect gas law ($p = n_{\mathrm{tot}}\,k\,T$), modified Saha ionization [3] and the M1 closure relation: $\mathsf{P}_r = \mathsf{D}\,E_r$ with $\mathsf{D} \equiv \chi$ the Eddington factor in 1D.

Table 1 Effective radiation source terms used in the three considered radiation regimes

	Full radiation transfer[a] (*)	Optically thin[b] regime (†)	Intermediate regimes
s_{E_r}	$\kappa_P\, \rho\, c\, (a\, T^4 - E_r)$	$n_e\, n_H\, \Lambda$	$\zeta\, s^*_{E_r} + (1-\zeta) s^\dagger_{Er}$
$\mathbf{s}_{\mathbf{M}_r}$	$-\kappa_R\, \rho\, c\, \mathbf{M}_r$	$\mathbf{0}^c$	$\mathbf{s}^*_{\mathbf{M}_r}$

[a]LTE
[b]Coronal
[c]Undefined source term
Λ: optically thin line cooling
$\zeta = \frac{1-\exp(3\tau_e)}{3\tau_e}$, $\tau_e = \kappa_R\, \rho\, L_c$

The coupling between the hydrodynamics (the three first equations) and the radiation field (the two last equations) is performed by the inclusion of the radiation energy and momentum (flux) density source terms (s_{E_r} and $\mathbf{s}_{\mathbf{M}_r}$ resp.) in the gas energy density and momentum source terms (s_e and $\mathbf{s}_m$ resp.) respectively. The Table 1 presents the expression of these source terms in different cases. The first column deals with the full radiation transfer expression near the LTE thermodynamical limit (higher densities, lower temperatures) based on our tabulated Planck (κ_P) and Rosseland (κ_R) mean opacities. The second column presents the optically thin coronal limit (higher temperatures, lower densities) based on the Λ cooling function provided by [12]. Due to the wide range of physical conditions of the system (from the stellar atmosphere up to the impacting flow), and in the absence on existing NLTE opacities, we had to use in this work an interpolation scheme between these two regimes. This model is based on the parameter ζ, the possibility for photons to escape the accretion column[1] sideways, then describing qualitatively multi-D effects for the radiation.

These equations are implemented in the ALE code AstroLabE [22]. Some of the 1D structures generated are then post-processed with the radiative transfer code Synspec [9].

3 QPO Perturbation

Our simplest model consists in an optically thin accretion flow impacting a rigid interface: a slab of post-shock material develops, then the cooling instability is triggered and the whole structure collapses, a new slab forms, and so on. The QPO characteristics obtained are consistent with [24]. After this test case, we have analyzed the influence of a dynamically-heated stellar atmosphere on an accretion flow, and then the effect of the radiation feedback on the whole structure.

[1]Of radius $L_c = 1000\,\text{km}$ in our simulations.

3.1 Dynamical Chromosphere and Optically Thin Flow

We model the stellar chromosphere by a plasma layer at hydro-radiative equilibrium, and perturbed by acoustic waves (10^8 erg/s at 17 mHz, see [18]) that degenerate into shocks in their upward propagation. One solar luminosity enters the simulation domain from the inner boundary. This toy model explains qualitatively the temperature profile of the solar chromosphere (see Fig. 1).
The (optically thin) accretion flow impacting such an atmosphere is hardly perturbed during most of the cycle. However, the acoustic heating induces the formation of a "new" slab before the "old" one ends to collapse. Figure 2 shows "old" and "new" slabs passing each other.

The net effect is that the QPO behavior of the post-shock material is perturbed *within* each fibril due to the chromosphere's dynamics. Measurement of a "pure" periodic signal due to the slab oscillation is therefore highly unlikely. The cycle duration in this model is slightly reduced (from 400 to 350 s).

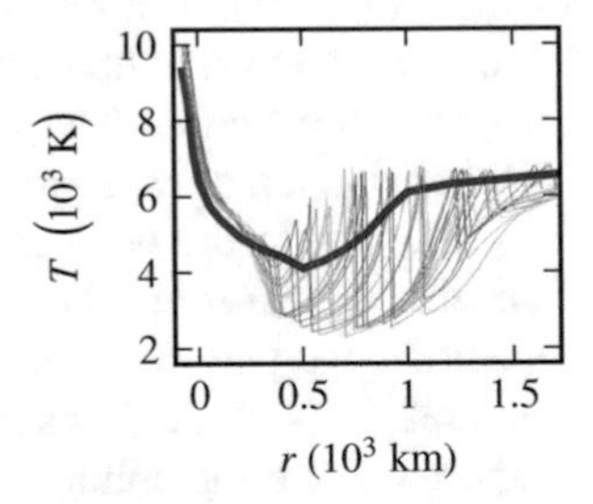

Fig. 1 Temperature snapshots of the simulated chromosphere (thin lines), and solar model (thick red line, [26]). (Adapted from [4])

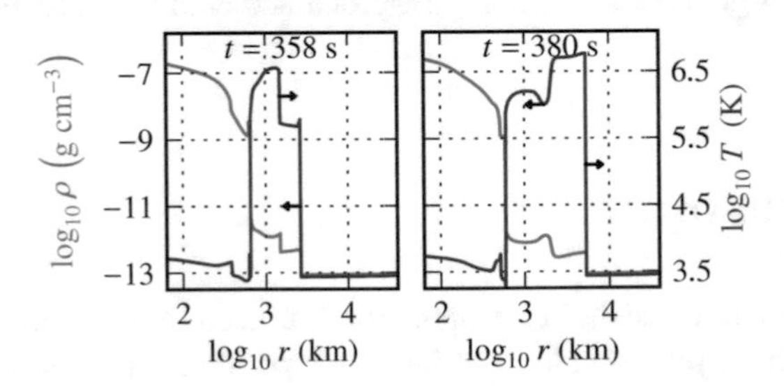

Fig. 2 Snapshots of the temperature (red) and density (green) profiles during a QPO cycle with dynamic chromosphere and coronal accretion flow. The arrows show the direction of propagation of the two main structures. (Adapted from [23])

3.2 Hybrid Radiation Transfer

In this setup, we switch off the acoustic heating of the chromosphere, and the whole computational domain (the atmosphere and the accretion flow) is modelled with the intermediate radiation source terms (cf. Table 1), enabling therefore radiation feedback between the stellar atmosphere and the accreted plasma. One solar luminosity still enters the simulation domain from the inner boundary.

Figure 3 shows four snapshots of the temperature and gas density during a QPO cycle with this setup. They reveal that the upper chromosphere is heated up and inflates due to the absorption of radiation emitted by the post-shock slab. The whole post-shock structure can thus be unburied (pushed upward over 2000 km). This effect comes along with a net reduction of the cycle duration (from 400 to ~150 s).

The results have been post-processed with the code Synspec [9], assuming LTE radiation transfer. Figure 4 (blue) presents the spectrum emerging from the accretion column. Three frequency bands can be distinguished: the Vis-IR band is dominated by the photospheric emission, the UV band is strongly absorbed by the accreted material, and the X-ray band contains the signature of the hot post-shock plasma. The red spectrum is the one obtained without considering the pre-shock medium. These two curves show the strong UV absorption by the cold pre-shock material.

The $\lambda-$integrated spectrum (between 2 and 27 Å), i.e. the radiation flux, is presented in Fig. 5 (blue curve). Although the computed flux is overestimated

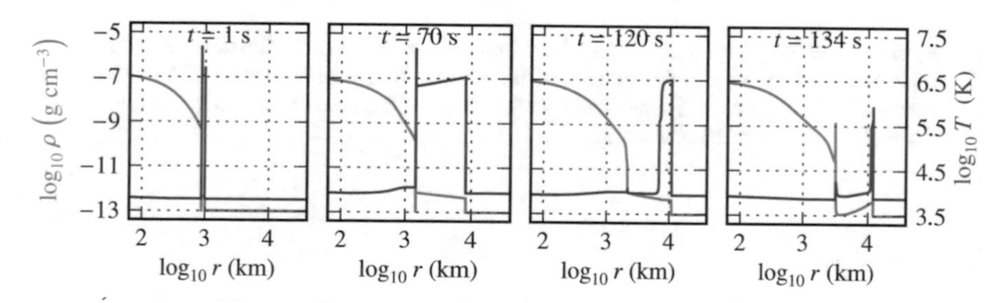

Fig. 3 Snapshots of the temperature (red) and density (green) profiles during a QPO cycle with hybrid radiation transfer. (Adapted from [23])

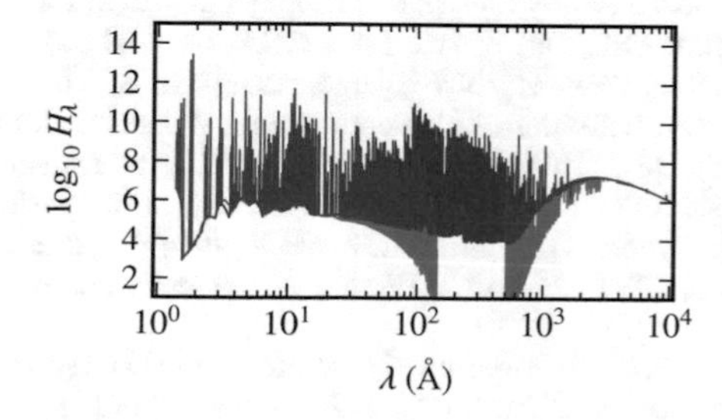

Fig. 4 Radiative Eddington flux H_λ (erg/cm^2/s/Å) emerging from the accretion column (blue) and from the slab (at $r \sim 10^4$ km, red) at $t = 70$ s (cf. Fig. 3). (Adapted from [23])

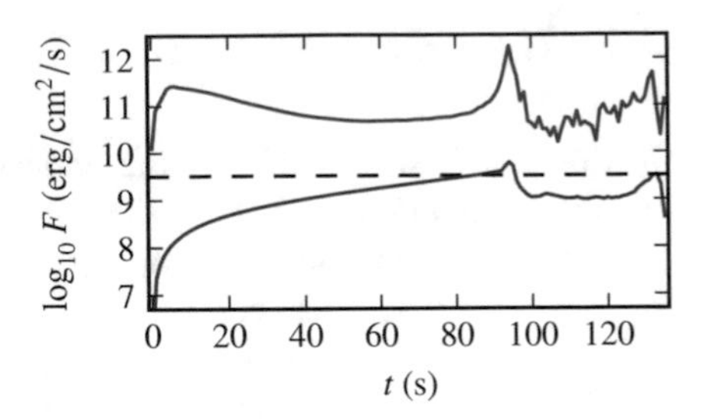

Fig. 5 Time-variation of the 2–27 Å integrated physical flux (blue) and of the optically thin post-shock emission (red) during a QPO cycle. The dashed line represents the incoming kinetic energy flux. (Adapted from [23])

(1.5 decades too high) with respect to the incoming kinetic energy flux (dashed line), it shows the same behavior than the flux derived from the optically thin cooling function [12] (red curve). Assuming a filling factor of 3%, this last one recovers the luminosity measured by [2] (1.3 10^{30} erg/s).

4 Prospects

Two main paths of investigations are considered to improve these results. We plan to use the multigroup approach to cover the specificity of each frequency band obtained in Sect. 3.2. This work shows the necessity to implement NLTE opacities and emissivities in the radiation equations: this work is under progress.

Acknowledgements These studies have been funded by the French "Programme National de Physique Stellaire" of INSU, the French Italian cooperation program PICS 6838 "Physics of Mass Accretion Processes in Young Stellar Objects", the Observatoire de Paris and the LABEX PLAS@PAR (ANR-11-IDEX-0004-02).

References

1. Bonito R, Orlando S, Argiroffi C, Miceli M, Peres G, Matsakos T, Stehlé C, Ibgui L (2014) Magnetohydrodynamic Modeling of the Accretion Shocks in Classical T Tauri Stars: the Role of Local Absorption in the X-Ray Emission. Astrophys. J. 795(2):L34–
2. Brickhouse NS, Cranmer SR, Dupree AK, Luna GJM, Wolk SJ (2010) A deep Chandrax-ray spectrum of the accreting young star TW Hydrae. Astrophys. J. 710(2):1835–1847
3. Brown JC (1973) On the ionisation of hydrogen in optical flares. Sol. Phys. 29(2):421–427
4. Chièze JP, de Sá L, Stehlé C (2013) Hydrodynamic modeling of accretion shocks on a star with radiative transport and a chromospheric model. EAS Publications Series 58:143–147
5. Colombo S, Orlando S, Peres G, Argiroffi C, Reale F (2016) Impacts of fragmented accretion streams onto classical T Tauri stars: UV and X-ray emission lines. Astron. Astrophys. 594:A93–
6. Costa G, Orlando S, Peres G, Argiroffi C, Bonito R (2017) Hydrodynamic modelling of accretion impacts in classical T Tauri stars: radiative heating of the pre-shock plasma. Astron. Astrophys. 597:A1

7. Drake JJ, Ratzlaff PW, Laming JM, Raymond JC (2009) An absence of X-ray accretion shock instability signatures in TW Hydrae. Astrophys. J. 703(2):1224–1229
8. Günther HM, Lewandowska N, Hundertmark MPG, Steinle H, Schmitt JHMM, Buckley D, Crawford S, O'Donoghue D, Vaisanen P (2010) The absence of sub-minute periodicity in classical T Tauri stars. Astron. Astrophys. 518:A54
9. Hubeny I, Lanz T (2017) A brief introductory guide to TLUSTY and SYNSPEC. eprint arXiv:170601859 pp –
10. Johns-Krull CM, Valenti JA, Hatzes AP, Kanaan A (1999) Spectropolarimetry of Magnetospheric Accretion on the Classical T Tauri Star BP Tauri. Astrophys. J. 510(1):L41–L44
11. Kastner JH, Huenemoerder DP, Schulz NS, Canizares CR, Weintraub DA (2002) Evidence for Accretion: High-Resolution X-Ray Spectroscopy of the Classical T Tauri Star TW Hydrae. Astrophys. J. 567(1):434–440
12. Kirienko AB (1993) Time-dependent radiative cooling of a hot, optically thin interstellar gas. Astron. Lett. 19:11–13
13. Lowrie RB, Mihalas D, Morel JE (2001) Comoving-frame radiation transport for nonrelativistic fluid velocities. JQSRT 69(3):291–304
14. Matsakos T, Chièze JP, Stehlé C, González M, Ibgui L, de Sá L, Lanz T, Orlando S, Bonito R, Argiroffi C, Reale F, Peres G (2013) YSO accretion shocks: magnetic, chromospheric or stochastic flow effects can suppress fluctuations of X-ray emission. Astron. Astrophys. 557:A69
15. Matsakos T, Chièze JP, Stehlé C, González M, Ibgui L, de Sá L, Lanz T, Orlando S, Bonito R, Argiroffi C, Reale F, Peres G (2014) 3D numerical modeling of YSO accretion shocks. EPJ Web of Conferences 64:04,003
16. Orlando S, Sacco GG, Argiroffi C, Reale F, Peres G, Maggio A (2010) X-ray emitting MHD accretion shocks in classical T Tauri stars. Astron. Astrophys. 510:A71
17. Orlando S, Bonito R, Argiroffi C, Reale F, Peres G, Miceli M, Matsakos T, Stehlé C, Ibgui L, de Sá L, Chièze JP, Lanz T (2013) Radiative accretion shocks along nonuniform stellar magnetic fields in classical T Tauri stars. Astron. Astrophys.
18. Rammacher W, Ulmschneider P (1992) Acoustic waves in the solar atmosphere. IX - Three minute pulsations driven by shock overtaking. Astron. Astrophys. 253:586–600
19. Romanova MM, Ustyugova GV, Koldoba AV, Wick JV, Lovelace RVE (2003) Three-dimensional Simulations of Disk Accretion to an Inclined Dipole. I. Magnetospheric Flows at Different Θ. Astrophys. J. 595(2):1009–1031
20. Romanova MM, Ustyugova GV, Koldoba AV, Lovelace RVE (2004) Three-dimensional Simulations of Disk Accretion to an Inclined Dipole. II. Hot Spots and Variability. Astrophys. J. 610(2):920–932
21. Romanova MM, Kulkarni AK, Lovelace RVE (2008) Unstable Disk Accretion onto Magnetized Stars: First Global Three-dimensional Magnetohydrodynamic Simulations. Astrophys. J. 673(2):L171–L174
22. de Sá L, Chièze JP, Stehlé C, Hubeny I, Delahaye F, Lanz T (2012) Hydrodynamic modeling of accretion shocks on a star with radiative transport and a chromospheric model. SF2A pp 309–312
23. de Sá L, Chièze JP, Stehlé C, Hubeny I, Lanz T, Delahaye F, Ibgui L (submitted) New insight on Young Stellar Objects accretion shocks. Astron. Astrophys.
24. Sacco GG, Argiroffi C, Orlando S, Maggio A, Peres G, Reale F (2008) X-ray emission from dense plasma in classical T Tauri stars: hydrodynamic modeling of the accretion shock. Astrophys. J. 491(2):L17–L20
25. Sacco GG, Orlando S, Argiroffi C, Maggio A, Peres G, Reale F, Curran RL (2010) On the observability of T Tauri accretion shocks in the X-ray band. Astron. Astrophys. 522:A55
26. Vernazza JE, Avrett EH, Loeser R (1973) Structure of the Solar Chromosphere. I - Basic Computations and Summary of the Results. Astrophys. J. 184:605

Radiation Magnetohydrodynamic Models and Spectral Signatures of Plasma Flows Accreting onto Classical T Tauri Stars

S. Colombo, L. Ibgui, R. Rodriguez, S. Orlando, M. González, C. Stehlé, and L. De Sá

1 Introduction

Classical T Tauri Stars (CTTSs) are young stars surrounded by a disk. According to the largely accepted magnetospheric accretion scenario [1], the disk extends internally until the, so called, truncation radius. Here the magnetic field is strong

S. Colombo (✉)
Dipartimento di Fisica & Chimica, Universitá degli Studi di Palermo, Palermo, Italy

LERMA, Observatoire de Paris, Sorbonne Université, École Normale Superieure, CNRS, Paris, France

Universidad de Las Palmas de Gran Canaria, Las Palmas, Spain
e-mail: salvatore.colombo@inaf.it

L. Ibgui · C. Stehlé
LERMA, Observatoire de Paris, Sorbonne Université, Ecole Normale Supérieure, CNRS, Paris, France
e-mail: laurent.ibgui@obspm.fr; chantal.stehle@obspm.fr

R. Rodriguez
Universidad de Las Palmas de Gran Canaria, Las Palmas, Spain
e-mail: ayaz@ing.uchile.cl

S. Orlando
INAF-Osservatorio Astronomico di Palermo "G.S. Vaiana", Palermo, Italy
e-mail: orlando@astropa.inaf.it

M. González
Paris Diderot University, AIM, CEA, CNRS, Paris, France
e-mail: matthias.gonzalez@cea.fr

L. De Sá
LERMA, Observatoire de Paris, Sorbonne Université, Université PSL, CNRS, Paris, France

CEA/DSM/IRFU/SAp-AIM, CEA Saclay, CNRS, Gif-sur-Yvette, France
e-mail: lionel.desa@obspm.fr

C. Sauty (ed.), *JET Simulations, Experiments, and Theory*, Astrophysics and Space Science Proceedings 55, https://doi.org/10.1007/978-3-030-14128-8_5

enough to dominate the plasma dynamics. The plasma is funneled by the magnetic field to form accretion columns that falls into the star.

Several lines of evidence support this idea, in particular accreting CTTSs show a soft X-ray (0.2–0.8 KeV) excess, with typical lines produced at temperature of $10^5 - 10^6$ K. This has been interpreted as due to the impacts of accreting material onto the stellar surface, at the impact region a shock is produced and dissipates the kinetic energy of the downfalling material, thereby heating up the plasma to temperature of few million degrees, producing X-ray emission [2, 3].

In the last 10 years the explanation of the soft X-ray excess in CTTSs in terms of accretion shocks was well supported by hydrodynamic (HD) and magnetohydrodynamic (MHD) models. Time-dependent one-dimensional (1D) models of radiative accretion shocks in CTTSs provided a first accurate description of the dynamics of the post-shock plasma [4, 5] In particular [5] proposed a 1D HD model of a continuous accretion flow, thus assuming the ratio between the thermal pressure and the magnetic pressure $\beta \ll 1$, impacting the chromosphere of a CTTS. Their model reproduces the main features of high spectral resolution X-ray observations of the CTTS MP Mus. More recently, 2D MHD models of accretion impacts have been studied [6–8]. 2D models allow to explore those cases where the $\beta \ll 1$ approximation cannot be applied and, therefore, the 1D approximation cannot be used. These models proved that the accretion dynamics strongly depends on the configuration and strength of the magnetic field. In particular, the atmosphere around the impact region can be perturbed by the accreting plasma.

All the previous models do not take into account the effects or radiative gains by the matter. The only published work where the radiation effects are considered is by Costa et al. [10]. This model is the first attempt to include the full radiative transfer (RT) effects in the framework of accretion impacts. Costa el al. [10] do not directly couple the RT effects with HD equations, but include them in an iterative way. More precisely they first solve the HD equations, then calculate the heating due to the RT, and then perform the simulation again including the calculated heating. This first approach could still prove that, in certain conditions, the radiation coming from the post-shock region may be absorbed by the unshocked material above in the accretion column. The absorption may heats up the unshocked accretion column at temperature between $10^4 - 10^6$ K.

In this work we propose the first simulation including the radiative transfer effects, in non-LTE regime coupled with the HD equations.

2 The Model

Our model describes an accretion column with uniform density of $10^{11}\,\mathrm{cm}^{-3}$ impacting onto the chromosphere of a CTTS. The accretion column is assumed to fall along the z-axis with an impact velocity of 500 km/s, and an initial temperature of 2×10^4 K. For the sake of simplicity, we assume plane parallel approximation,

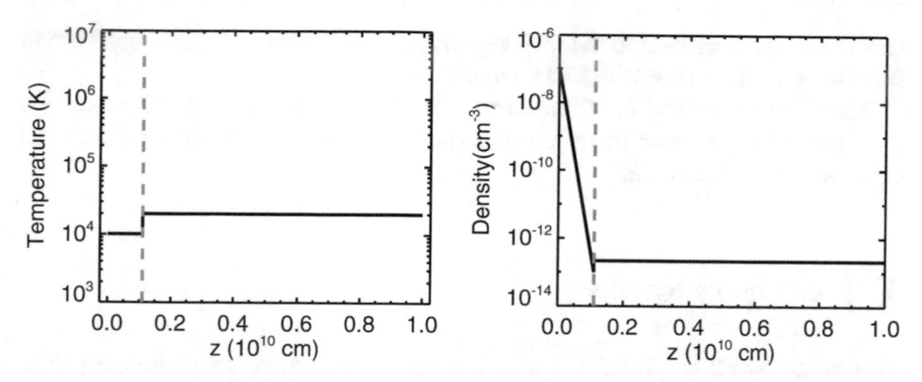

Fig. 1 Initial profiles of temperature (on the left) and density (on the right) for the simulation. The dotted vertical lines indicates the initial position of the chromosphere

and follow the evolution of the internal region of the accretion column. It is the same as assuming that in all the domain $\beta \ll 1$.

Initially, the accretion column, which is unshocked, is placed just above an idealized chromosphere, which is assumed to be at uniform temperature at 10^4 K. Figure 1 shows the initial conditions. The model solves the radiative hydrodynamics (RHD) equations: conservation of mass, momentum, total plasma energy (ϵ) and comoving-frame radiation energy (E), taking into account the gravity from the central star, the thermal conduction and the radiative heating and radiative losses. The total radiative properties (Plank mean opacity k_P, Rosseland mean opacity k_R, and radiative losses L) are calculated in the NLTE regime [12]. The set of RHD equations that we solve, under the flux-limited diffusion approximation, is

$$\frac{\partial \rho}{\partial t} + \nabla \cdot (\rho \mathbf{u}) = 0 \tag{1}$$

$$\frac{\partial \rho \mathbf{u}}{\partial t} + \nabla \cdot (\rho \mathbf{u} \times \mathbf{u}) + \nabla p = \rho \mathbf{g} + \frac{\rho k_R}{c} \mathbf{F} \tag{2}$$

$$\frac{\partial \epsilon}{\partial t} + \nabla \cdot [(\epsilon + p)\mathbf{u}] = \rho \mathbf{u} \mathbf{g} + \nabla \cdot F_c - L + k_P \rho c E \tag{3}$$

$$\frac{\partial E}{\partial t} + \nabla \cdot \mathbf{F} = L - k_P \rho c E \tag{4}$$

$$p = \rho \frac{k_B T}{\mu m_H} \qquad \mathbf{F} = -\lambda \frac{c}{k_R \rho} \nabla E \tag{5}$$

where ρ is the density, $\mathbf{u}$ the velocity, p the gas pressure, $\mathbf{g}$ the gravity, F_c the thermal conduction, c the speed of light, $\mathbf{F}$ the comoving-frame radiation flux, and λ the flux limiter. The equation are solved in a Cartesian coordinates system (x,y,z).

The calculation were performed using PLUTO v4.0 [9], a modular, Godunov-type code for astrophysical plasmas. PLUTO was coupled with a RT module, which was originally restrained to the LTE regime [11], and which we have upgraded in order to take into account the NLTE conditions.

The domain consists of a 3D uniform grid with only 3 points for x and y-axes and 8192 points for the z-axis, this grid was chosen as a trade-off between computational cost and spatial resolution.

3 Preliminary Results

This is still work in progress, hence the results presented are preliminary. The evolution of the system is shown in Fig. 2.

Initially, the accretion column is located just above the chromosphere.

The density map (Fig. 2 left) shows that, initially, the accretion column sinks into the chromosphere. It stops when the thermal pressure in the chromosphere equals the ram-pressure of the stream. At this point, a shock propagates through the accretion column forming a post-shock region (light blue in Fig. 2 left and dark red in Fig. 2 right). The post-shock regions has a transient phase (between 100 and 300 s), where the accretion column is still sinking into the chromosphere. During the transient phase the post-shock region extends up to $\approx 5 \times 10^8$ cm above the impact region. After the transition phase the post-shock region increases, reaching a maximum value of $\approx 3 \times 10^9$ cm.

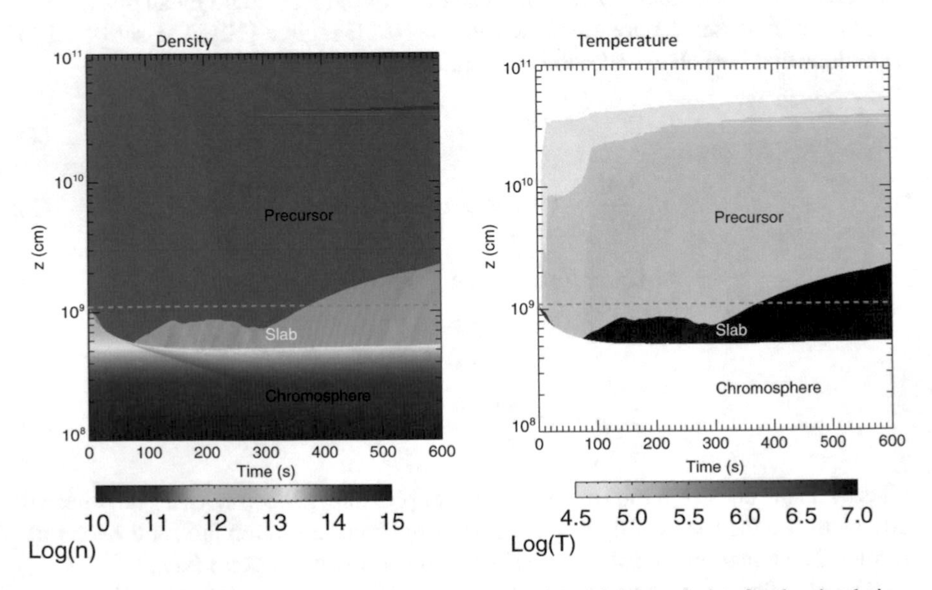

Fig. 2 Time-space plot of the density (left) and temperature (right) evolution for the simulation. The spatial extent of the shock lies in the vertical direction. The horizontal direction indicates the time. The dashed grey line indicates the initial position of the chromosphere

Moreover, the temperature map shows that the shock heats up the plasma, forming a post-shock region at 10^6 K. This region strongly radiates in UV and X-ray bands. At these wavelengths the unshocked material above is optically thick and absorbs part of the radiation. As a result, a precursor region develops. The precursor is composed of two different regions, the first one with an extension of $\approx 2 \times 10^{10}$ cm and a temperature of $\approx 5 \times 10^5$ K, the latter with a maximum extension of $\approx 4 \times 10^{10}$ cm and a temperature of $\approx 5 \times 10^4$ K. It is important to stress that, in this simulation, we assume a plane parallel geometry, which means that we consider an accretion stream with an infinite horizontal extension.

We can conclude that, RHD simulations that include, for the first time, the radiation effects in NLTE regime, suggest that:

1. Part of the UV and X-ray radiation produced by the accretion shock in CTTSs is absorbed by the upstream part of the accretion column
2. The effect of the absorption is to heat up the plasma at temperature of 10^5 K, forming a precursor region that has to be considered as a new source of UV emission in the framework of accretion phenomena

Acknowledgements PLUTO is developed at the Turin Astronomical Observatory in collaboration with the Department of Physics of Turin University. We acknowledge the INAF – Osservatorio Astronomico di Palermo, for the availability of high performance computing resources and support. This work was supported by the Programme National de Physique Stellaire (PNPS) of CNRS/INSU co-funded by CEA and CNES. This work has been done within the LABEX Plas@par project, and received financial state aid managed by the Agence Nationale de la Recherche (ANR), as part of the programme "Investissements d'avenir" under the reference ANR-11-IDEX-0004-02.

References

1. Koenigl, A.: Disk accretion onto magnetic T Tauri stars. ApJ. **370**, L39-L43 (1991)
2. Kastner, Joel H.; Huenemoerder, David P.; Schulz, Norbert S.; Canizares, Claude R.; Weintraub, David A.: Evidence for Accretion: High-Resolution X-Ray Spectroscopy of the Classical T Tauri Star TW Hydrae. ApJ. **567**, 434–440 (2002)
3. C. Argiroffi, A. Maggio and G. Peres: X-ray emission from MP Muscae: an old classical T Tauri star. ApJ. **465**, L5-L8 (2007)
4. Koldoba, A. V.; Ustyugova, G. V.; Romanova, M. M.; Lovelace, R. V. E.: Oscillations of magnetohydrodynamic shock waves on the surfaces of T Tauri stars. ApJ. **388**, 357–366 (2008)
5. G.G. Sacco, C. Argiroffi, S. Orlando, A. Maggio, G. Peres and F. Reale: X-ray emission from MP Muscae: an old classical T Tauri star. ApJ. **491**, L17-L20 (2008)
6. S. Orlando G.G. Sacco, C. Argiroffi,F. Reale, G. Peres and A. Maggio:X-ray emitting MHD accretion shocks in classical T Tauri stars Case for moderate to high plasma-β values. ApJ. **510**, A71 (2010)
7. S. Orlando, R. Bonito, C. Argiroffi, F. Reale, G. Peres, M. Miceli, T. Matsakos, C. Stehlé, L. Ibgui, L. de Sa, J. P. Chièze and T. Lanz: Radiative accretion shocks along nonuniform stellar magnetic fields in classical T Tauri stars. ApJ. **559**, A127 (2013)
8. T. Matsakos, J.-P. Chièze, C. Stehlè, M. González, L. Ibgui, L. de Sá, T. Lanz, S. Orlando, R. Bonito, C. Argiroffi, F. Reale and G. Peres: YSO accretion shocks: magnetic, chromospheric or stochastic flow effects can suppress fluctuations of X-ray emission **557**, A69 (2013)

9. Mignone, A.; Bodo, G.; Massaglia, S.; Matsakos, T.; Tesileanu, O.; Zanni, C.; Ferrari, A.: PLUTO: A Numerical Code for Computational Astrophysics. ApJ. **170**, 228–242 (2007)
10. G. Costa, S. Orlando, G. Peres, C. Argiroffi, and R. Bonito: Hydrodynamic modelling of accretion impacts in classical T Tauri stars: radiative heating of the pre-shock plasma. ApJ. **597**, A1 (2017)
11. S.M. Kolb, M. Stute, W. Kley, and A. Mignone: Radiation hydrodynamics integrated in the PLUTO code. ApJ. **559**,A80 (2013)
12. Rodríguez, R., Espinosa, G., and Gil, J. M.: Phys. Rev. E. **98**, 033213 (2018)

Mass Accretion Impacts in Classical T Tauri Stars: A Multi-disciplinary Approach

S. Orlando, C. Argiroffi, R. Bonito, S. Colombo, G. Peres, F. Reale, M. Miceli, L. Ibgui, C. Stehlé, and T. Matsakos

1 Introduction

According to the magnetospheric accretion scenario, young low-mass stars are surrounded by circumstellar disks with which they interact in a complex fashion, with accretion of mass and ejection of collimated outflows. The accretion process is regulated by the stellar magnetic field which disrupts the inner part of the disk at a distance of a few stellar radii (the truncation radius) and guides the disk's material toward the central star. The impact of this material onto the stellar surface is expected to generate a shock which propagates through the accretion column and heats the downfalling material up to temperatures of million degrees. The impacts, therefore, are expected to generate X-ray emission.

S. Orlando (✉) · R. Bonito
INAF – Osservatorio Astronomico di Palermo, Palermo, Italy
e-mail: salvatore.orlando@inaf.it; rosaria.bonito@inaf.it

C. Argiroffi · G. Peres · F. Reale · M. Miceli
Dip. di Fisica e Chimica, Università di Palermo, Sicily, Italy

INAF – Osservatorio Astronomico di Palermo, Palermo, Italy
e-mail: costanza.argiroffi@unipa.it; giovanni.peres@unipa.it; fabio.reale@unipa.it; marco.miceli@unipa.it

S. Colombo
INAF – Osservatorio Astronomico di Palermo, Palermo, Italy

LERMA, Sorbonne Université, Observatoire de Paris, Université PSL, CNRS, Paris, France
e-mail: salvatore.colombo@inaf.it

L. Ibgui · C. Stehlé · T. Matsakos
LERMA, Sorbonne Université, Observatoire de Paris, Université PSL, CNRS, Paris, France
e-mail: laurent.ibgui@obspm.fr; chantal.stehle@obspm.fr; titos.matsakos@obspm.fr

C. Sauty (ed.), *JET Simulations, Experiments, and Theory*, Astrophysics and Space Science Proceedings 55, https://doi.org/10.1007/978-3-030-14128-8_6

In fact, high resolution X-ray observations of young stars accreting material from their circumstellar disks (e.g. TW Hya, BP Tau, V4046 Sgr, MP Mus and RU Lupi) have revealed X-ray emission from plasma at $T \approx 2$–5 MK, which is denser than $n_H = 10^{11}\,\mathrm{cm}^{-3}$ [9]. This soft X-ray emission component could be produced by the material accreting onto the star surface, flowing along the magnetic field lines of the nearly dipolar stellar magnetosphere, and heated to temperatures of few MK by a shock at the base of the accretion column [3, 11].

In the last years several models have been proposed to describe the impacts of accreting material onto the surface of classical T Tauri stars (CTTSs). They provide a convincing theoretical support and a plausible global picture of the phenomenon at work. However some fundamental aspects of accretion impacts still need to be clarified: the observed X-ray luminosity produced by plasma heated in accretion shocks is largely below the predicted value [6]; the observed coronal activity seems to be influenced by accretion but it is not clear why and how [8, 14]; UV emission lines exhibit complex profiles and large doppler shifts [1]. In this contribution, we briefly review some of the achievements obtained by our group by exploiting a multi-disciplinary approach based on the analysis of multi-dimensional magnetohydrodynamic simulations, multi-wavelength observations, and laboratory experiments of accretion impacts occurring onto the surface of CTTSs.

2 Structure and Evolution of Shock-Heated Plasma in Accretion Impacts

The first numerical models describing the impact of an accretion column onto the surface of a CTTS were one-dimensional (1D) and they have shown that the continuous impact of an accretion column onto the stellar surface leads to the formation of a dense and hot slab of plasma that undergoes sandpile quasi-periodic (QPOs) oscillations driven by catastrophic cooling [10, 19, 20]. The major achievement of these models has been to demonstrate that the main features of high spectral resolution X-ray observations of CTTSs can be reproduced by accretion impacts and originate from post-shock plasma [19]. These models however are justified if the plasma has a $\beta \ll 1$ (where β = gas pressure / magnetic pressure) in the shock-heated material: in this conditions the plasma is assumed to move and to transport energy only along magnetic field lines.

The stability and dynamics of accretion impacts in cases where the low-β approximation cannot be applied have been investigated through two-dimensional (2D) magnetohydrodynamic (MHD) models [4, 12, 13, 15]. These models have shown that, in general, QPOs cannot be observed (according to the absence of periodic X-ray modulation due to shock oscillations found in observations; [7]) due to heavy dumping by the magnetic field [13] or to perturbation of the stream induced by the post-shock plasma itself [4, 12, 15]. The atmosphere around the impact region of the stream can be strongly perturbed (depending on the plasma β), leading to important leaks at the border of the main stream [15].

The 2D MHD models have also shown that the strength and configuration of the magnetic field play a crucial role in determining the dynamics and evolution of the post-shock plasma. In the case of weak magnetic fields (plasma $\beta \gtrsim 1$ in the post-shock region), a large component of **B** may develop perpendicular to the stream at the base of the accretion column. This component limits the sinking of the shocked plasma into the chromosphere [13, 15]. For strong magnetic fields ($\beta < 1$ in the post-shock region close to the chromosphere), the field configuration determines the position of the shock and its stand-off height [13]. If the field is strongly tapered close to the chromosphere, an oblique shock may form well above the stellar surface at the height where the plasma $\beta \approx 1$.

The above models have allowed to investigate the structure, stability, and location of the post-shock plasma. However they do not explain the evidence that a significant amount of plasma at 10^5 K is produced in the accretion process. Also they are not able to reproduce UV observations which show the C IV doublet at 1550 Å with a complex profile described by two Gaussian components with different speeds and widths [1]: a narrow component redshifted at speeds of $\approx 30\,\mathrm{km\,s^{-1}}$ and a broader component centered at $\approx 120\,\mathrm{km\,s^{-1}}$ and with the redshifted wing reaching $\approx 400\,\mathrm{km\,s^{-1}}$.

Surprisingly solar observations suggested a way to explain the origin of the observed asymmetries and redshifts of UV emission lines in CTTSs. After a violent eruption occurred on 2011 June 7, dense plasma fragments were observed to fall back on the surface of the Sun, producing brightenings in X-rays after the impacts. These impacts have been shown to reproduce on the small scale accretion impacts onto CTTSs [17]. The hydrodynamic modeling of the impacts observed in the Sun and the synthesis of emission in UV and X-ray bands from the models have shown that UV emission may originate from the shocked front shell of the still downfalling fragments [16]. In this case a broad redshifted component in UV lines is produced up to speeds around $\approx 400\,\mathrm{km\,s^{-1}}$ which is consistent with UV observations of CTTSs.

The hypothesis that the accreting material can be fragmented or clumpy has been challenged by an MHD model which describes an accretion column consisting of several high density blobs which impact onto the chromosphere of a CTTS [4]. The model has shown that the impacts of the blobs produce shocked upflows which, possibly, hit and shock the subsequent downfalling fragments. As a result, several shocked plasma components with different downfalling velocities are present altogether, leading to profiles of C IV lines remarkably similar to those observed in CTTSs. The fragmentation of the accreting material, therefore, may explain in a natural way the origin of asymmetries and redshifts of UV emission lines in CTTSs.

3 Effect of Optically Thick Plasma Around Impact Regions

The analysis of observations in different bands has shown the evidence that the mass accretion rates derived from X-rays are significantly lower than those derived from other spectral bands (UV/optical/NIR observations; [6]). Early 1D models of accretion impacts have suggested a reason for this discrepancy. They pointed out the importance of absorption from the optically thick material of the chromosphere in the post-shock plasma components that produce observable emission in the X-ray band [20]. In addition 2D MHD models have shown that absorption of X-ray emission may originate from the presence of optically thick material located around the impact regions. In the case of plasma $\beta \gtrsim 1$ in the post-shock region, an envelope of dense and cold chromospheric material may develop around the shocked column. This envelope is expected to determine a significant absorption of the X-ray emission arising from the post-shock plasma [13].

The synthesis of X-ray emission from 2D MHD models has shown that, if the effects of local absorption are taken into account, the X-ray fluxes inferred from the emerging spectra are lower than expected because of the complex local absorption by the optically thick material of the chromosphere and of the unperturbed stream [2]. The first attempt to investigate in more detail the effects of radiative transfer during accretion impacts in CTTSs was performed through a 1D hydrodynamic model [5]. This model has shown that the dense and cold plasma of the pre-shock accretion column is gradually heated up to a few 10^5 K due to irradiation of X-rays arising from the shocked plasma at the impact region. As a result, a region of radiatively heated gas (a precursor) forms in the unshocked accretion column and contributes significantly to UV emission. In such a way, the model naturally reproduces the luminosity of UV emission lines correlated to accretion and shows that most of the UV emission originates from the precursor.

The hypothesis of absorption of X-ray emission in accretion impacts has been strongly supported by the analysis of solar observations and laboratory experiments. Observations of bright hot impacts by erupted fragments falling back on the Sun have shown that the X-ray emission produced in the impacts is heavily absorbed by optically thick plasma, possibly explaining the discrepancy between mass accretion rates derived from X-ray and optical observations [17]. Scaled laboratory experiments of collimated plasma accretion onto a solid in the presence of a magnetic field were able to track, with spatial and temporal resolution, the dynamics of the system and simultaneously measured multi-band emissions [18]. The experiments have shown that optically thick material due to the perturbation of the chromosphere may envelope the impact region and determine a partial absorption of the X-ray emission. Once again, this finding supports absorption as possible reason reconciling current discrepancies between mass accretion rates derived from X-ray and optical observations, respectively.

4 Summary and Conclusions

We summarized some of the recent achievements in the study of accretion impacts in CTTSs obtained by adopting a multi-disciplinary approach. The MHD models provide a plausible global picture of the phenomenon at work. They were able to investigate the effects of stellar magnetic field in determining the structure, stability, and location of the shocked plasma. In particular they have suggested that the distribution of dense and cold material around the hot slab may lead to significant absorption of X-ray emission arising from the post-shock plasma. The structure of the hot slab can be largely affected by chromospheric or stochastic flow effects. The analysis of observability of accretion shocks in UV and X-ray bands has shown that the observed asymmetric and redshifted line profiles can be explained by fragmented or clumpy accretion columns. The laboratory experiments can provide important complementary information on the physics of accretion impacts. For instance, it was possible to prove that heavy absorption of X-ray emission by optically thick material is present around impact regions. The evidence of absorption of X-ray emission may reconcile current discrepancies between mass accretion rates derived from X-ray and optical observations. Finally, it was proved that solar observations can provide a powerful template for stellar accretion. Thanks to these observations it was possible to prove the importance of local absorption in suppressing the X-ray emission from the post-shock plasma and to find a way to explain the origin of the observed asymmetries and redshifts of UV emission lines in CTTSs.

Acknowledgements The PLUTO code, used in this work, is developed at the Turin Astronomical Observatory in collaboration with the Department of General Physics of Turin University and the SCAI Department of CINECA. These studies have been partially funded by the French Italian cooperation program PICS 6838 "Physics of Mass Accretion Processes in Young Stellar Objects", and the LABEX PLAS@PAR (ANR-11-IDEX-0004-02).

References

1. Ardila, D.R., Herczeg, G.J., Gregory, S.G., Ingleby, L., France, K., Brown, A., Edwards, S., Johns-Krull, C., Linsky, J.L., Yang, H., Valenti, J.A., Abgrall, H., Alexander, R.D., Bergin, E., Bethell, T., Brown, J.M., Calvet, N., Espaillat, C., Hillenbrand, L.A., Hussain, G., Roueff, E., Schindhelm, E.R., Walter, F.M. 2013, ApJS, 207, 1
2. Bonito, R., Orlando, S., Argiroffi, C., Miceli, M., Peres, G., Matsakos, T., Stehle, C., Ibgui, L. 2014, ApJ, 795, L34
3. Calvet, N., Gullbring, E. 1998, ApJ, 509, 802
4. Colombo, S., Orlando, S., Peres, G., Argiroffi, C., Reale, F. 2016, A&A, 594, A93
5. Costa, G., Orlando, S., Peres, G., Argiroffi, C., Bonito, R. 2017, A&A, 597, A1
6. Curran, R.L., Argiroffi, C., Sacco, G.G., Orlando, S., Peres, G., Reale, F., Maggio, A. 2011, A&A, 526, A104
7. Drake, J.J., Ratzlaff, P.W., Laming, J.M., Raymond, J. 2009, ApJ, 703, 1224
8. Flaccomio, E., Damiani, F., Micela, G., Sciortino, S., Harnden Jr., F.R., Murray, S.S., Wolk, S.J. 2003, ApJ, 582, 398

9. Kastner, J.H., Huenemoerder, D.P., Schulz, N.S., Canizares, C.R., Weintraub, D.A. 2002, ApJ, 567, 434
10. Koldoba, A.V., Ustyugova, G.V., Romanova, M.M., Lovelace, R.V.E. 2008, MNRAS, 388, 357
11. Lamzin, S.A. 1998, Astronomy Reports, 42, 322
12. Matsakos, T., Chièze, J.P., Stehlé, C., González, M., Ibgui, L., de Sá, L., Lanz, T., Orlando, S., Bonito, R., Argiroffi, C., Reale, F., Peres, G. 2013, A&A, 557, A69
13. Orlando, S., Bonito, R., Argiroffi, C., Reale, F., Peres, G., Miceli, M., Matsakos, T., Stehlé, C., Ibgui, L., de Sa, L., Chièze, J.P., Lanz, T. 2013, A&A, 559, A127
14. Orlando, S., Reale, F., Peres, G., Mignone, A. 2011, MNRAS, 415, 3380
15. Orlando, S., Sacco, G.G., Argiroffi, C., Reale, F., Peres, G., Maggio, A. 2010, A&A, 510, A71
16. Reale, F., Orlando, S., Testa, P., Landi, E., Schrijver, C.J. 2014, ApJ, 797, L5
17. Reale, F., Orlando, S., Testa, P., Peres, G., Landi, E., Schrijver, C.J. 2013, Science, 341, 251
18. Revet, G., Chen, S.N., Bonito, R., Khiar, B., Filippov, E., Argiroffi, C., Higginson, D.P., Orlando, S., Béard, J., Blecher, M., Borghesi, M., Burdonov, K., Khaghani, D., Naughton, K., Pépin, H., Portugall, O., Riquier, R., Rodriguez, R., Ryazantsev, S.N., Skobelev, I.Y., Soloviev, A., Willi, O., Pikuz, S., Ciardi, A., Fuchs, J. 2017, Science Advances, 3, e1700982
19. Sacco, G.G., Argiroffi, C., Orlando, S., Maggio, A., Peres, G., Reale, F. 2008, A&A, 491, L17
20. Sacco, G.G., Orlando, S., Argiroffi, C., Maggio, A., Peres, G., Reale, F., Curran, R.L. 2010, A&A, 522, A55

High Energy Emission from Shocks Due to Jets and Accretion in Young Stars with Disks: Combining Observations, Numerical Models, and Laboratory Experiments

Rosaria Bonito, Costanza Argiroffi, Salvatore Orlando, Marco Miceli, Julien Fuchs, and Andrea Ciardi

1 Introduction

Accretion of material from the surrounding medium onto the stellar surface and ejection of material in form of jets are processes related to each other in young stars. These phenomena are observed during all the early phases of evolution of young stellar objects (YSO), from embedded class 0 and class I YSOs to class II or Classical T Tauri Stars (CTTS).

Jets from YSOs have been detected in all the wavelength, from radio, IR, optical, and also at high energy in the X-ray band, showing a knotty structure along the jet propagation. A complex morphology with also a detectable proper motion of the internal knots has been observed in X-rays by Favata, Bonito et al. [1] for the most luminous and close jet HH 154 in Taurus. Multiple observations with Chandra [2] reveal a stationary X-ray source at the base of the jet and an elongated

R. Bonito (✉) · S. Orlando
INAF – Osservatorio Astronomico di Palermo, Palermo, Italy
e-mail: rosaria.bonito@inaf.it; salvatore.orlando@inaf.it

C. Argiroffi · M. Miceli
Dipartimento di Fisica e Chimica, Università degli Studi di Palermo, Palermo, Italy
e-mail: costanza.argiroffi@inaf.it; marco.miceli@inaf.it

J. Fuchs
LULI CNRS, Ecole Polytechnique, Sorbonne Universités, CEA, UPME, Palaiseau cedex, France

Sorbonne Universités, Palaiseau cedex, France
e-mail: julien.fuchs@polytechnique.edu

A. Ciardi
Sorbonne Universités, Observatoire de Paris, PSL Research University, CNRS, LERMA, Paris, France
e-mail: andrea.ciardi@obspm.fr

C. Sauty (ed.), *JET Simulations, Experiments, and Theory*,
Astrophysics and Space Science Proceedings 55,
https://doi.org/10.1007/978-3-030-14128-8_7

structure consistent with a jet velocity of 500 km/s, in good agreement with the X-ray properties (temperature and luminosity) derived from the fit [1, 3].

The related opposite process of accretion in YSOs, generates shocks at the stellar surface with free-fall velocity of $\approx$400 km/s. The corresponding post-shock velocity expected is $\sim$100 km/s, leading to temperature of about 1−3 MK, therefore soft X-ray emission can be detected using high resolution instruments as XMM/Newton and Chandra.

The study of X-ray emission from accretion/ejection in YSOs can be important as they can alter the physical and chemical conditions of the circumstellar disks, having also an effect on their lifetime and possibly inhibiting the formation of exo-planetary systems.

2 Jets

In order to explain the stationary and moving components of the X-ray source detected in the well studied jet HH 154, we developed several numerical models that can reproduce the knotty structure along the jet axis, as the result of a pulsed scenario, also consistent with the hypothesis that the related accretion process should be episodic [4, 5], and the X-ray emission at the base of the jet, considering the formation of a shock diamond due to a nozzle ([2]; see also [6] for a magnetized jet).

Our models succeeded in reproducing the observations of HH 154 and DG Tau as well in X-rays. The nozzle proposed in [2] allowed us also to derive the strength of the magnetic field, a parameter which is difficult to directly observe in such embedded regions (being the progenitor star of HH 154 a class 0/I binary system), therefore our models have been used as probe of the launching/collimation region in obscured systems.

In [7], we explored a new interdisciplinary approach that combines astrophysical observations, numerical models, and laboratory experiments to investigate stellar jets. Starting from laser experiments reproducing a scaled version of astrophysical jets, we performed numerical simulations tuned on the laboratory experiments and we developed models of astrophysical jets. Then we synthesized the X-ray emission maps and compared our results with Chandra observations of the best studied jet, HH 154. The good agreement with the data make this approach promising for future comparison of astrophysical objects and laboratory scaled versions.

3 Accretion

Prompted by the promising results obtained for the comparison between laboratory experiments of jets and observations, mainly in the X-ray band, we developed an analogous experiment to reproduce the accretion process in young stars. We investigated the shock emission due to the accretion of material in young stars to solve some of the open issues still debated in the literature: the role of the local absorption, due to the surrounding material; the detectability of the Doppler shift, due to the motion of the plasma along the line-of-sight (LoS); the origin of the soft X-ray emission; the relation between the UV and the X-ray emission from the post-shock region. To this end, we developed numerical models exploring the magnetic field topology and intensity [8], and the effects of varying the local absorption and the inclination of the system with respect to the LoS [9]. We performed magnetohydrodynamic (MHD) simulations using the PLUTO code [10] considering a realistic radial profile of the accretion stream (as also suggested by [11]). We synthesized the UV and X-ray emission of the post-shock region, also exploring the detectability of the Doppler shift in the emission lines [12]. We have found that different wavelength emission suffers different absorption due to the surrounding material (the unperturbed accretion stream itself as well as the perturbed chromosphere) and as a result from the multi-band analysis we derive that different emission arises from different regions.

Our model's results in the UV lead to values lower than observations, while the X-rays model and observations are in good agreement. We can suggest that the reason could be due to: different streams that can emit in different bands; the density that can be variable in time; the contribution to UV that can be from both the post- and the pre-shock regions.

We have also investigated the detectability of the Doppler shift, by synthesizing the OVIII emission line profile predicted from our model and by comparing this result with the observations of TW Hya [13].

The redshift due to the plasma motion of infalling onto the stellar surface is not just predicted but has been actually observed for the first time. The value of this redshift (both predicted and observed) is small but significant and the result is interesting for several reasons: (1) the observations are in good agreement with the model's prediction, verifying the model itself; (2) the observed Doppler shift is the first detection; (3) the shift is actually a redshift, thus confirming that this X-ray emission originates in the post-shock region, and not in coronal structures; (4) we can infer indirect information about the geometry of the system: as discussed in [13], as we expect speed of $\sim$100 km/s, but derived $\approx$35 km/s, we suggest that the accretion stream we observed in TW Hya is at low latitude.

We have explored the impact of an accretion stream on the stellar surface as we have previously made for jets, combining observations, numerical models, and laboratory experiments. In [14], we have investigated this issue, obtaining a good description of the astrophysical problem of accretion shocks in young stars with a scaled version in the laboratory.

4 Future Perspectives

As we are pushing to the limits the capabilities of currently available instruments, as Chandra in X-rays, we need future instruments to further investigate accretion and ejection processes in young stars. Future X-ray mission Athena X-ray Observatory, with higher effective area, will allow us to improve the statistics of objects to be observed, with the consequence of exploring different ages, masses, and geometric configurations to study the relation between these properties and the accretion shock emission.

Our work on young stars with accretion and ejection processes includes a multi-band investigation, and in the next future we will also use the Large Synoptic Survey Telescope (LSST) for a complete characterization of such complex systems. Accretion (and related ejection) processes from young stars show strong variability in different wavelength bands. These systems can be studied both during the quiescent phase as well as during eruptive bursts. The accretion processes from circumstellar disks in young stars can be investigated in the u-band and blue bands exploring both their properties and variability: the variability of PMS accretion can be performed taking advantage of LSST colors. A preparatory study of the most interesting individual objects or of the most promising clusters (i.e. those who show high frequency of accretors) is under way using a multi-band and multi-approach technique, taking advantage of both observations (imaging, photometry, spectral data in synergy with current optical spectroscopic surveys, X-ray images, UV/optical/NIR data, etc.), models, and laser experiments [7, 14]. This is also interesting for the impact that the characterization of the accretion/outflow process can have in the context of exoplanetary systems formation and evolution and in the study of exoplanet population in clusters.

References

1. Favata, Bonito et al. 2006, A&A, 450, 17
2. Bonito et al. 2011, APJ, 737, 54
3. Bonito et al. 2007, A&A, 462, 645
4. Bonito et al. 2010a, A&A, 511, 42
5. Bonito et al. 2010b, A&A, 517, 68
6. Ustamujic et al. 2016, A&A, 596, 99
7. Albertazzi et al. 2014, Science, 346, 325
8. Orlando, Bonito et al. 2013, A&A, 559, 127
9. Bonito et al. 2014, APJ, 795, 34
10. Mignone et al. 2007, APJS, 170, 228
11. Romanova et al. 2004, APJ, 610, 920
12. Bonito et al. 2018, in prep
13. Argiroffi et al. 2017, A&A, 607, 14
14. Revet et al. 2017, Science Adv., 3(11)

On the Origin of the X-Ray Emission in Protostellar Jets Close to the Launching Site

S. Ustamujic, S. Orlando, R. Bonito, M. Miceli, and A. I. Gómez de Castro

1 Introduction

Jets are a fundamental part of the formation and evolution processes of young stars, being responsible for the removal of mass and angular momentum from the star-disk system. They form very complex structures as a result of the variable nature of the ejection mechanism and the subsequent interactions of the ejected material with the surrounding gas, emitting in different wavelength bands. In particular, high energy X-ray emission detected in a number of jets close to the launching site could have a relevant effect in the whole system, ionizing large parts of the circumstellar environment (for a review, see [13]).

X-ray observations showed evidence of faint X-ray emitting sources located at the base of the jet (e.g. [8, 14, 16]). Multi-epoch observations suggested the presence of stationary X-ray emitting sources in some cases, e.g., in HH 154 [1, 7, 15] and DG Tau [9–11]. In this work we study the origin, variability and detectability of X-ray emitting sources in protostellar jets located close to the launching site.

S. Ustamujic (✉) · S. Orlando
INAF-Osservatorio Astronomico di Palermo, Palermo, Italy
e-mail: sabina.ustamujic@inaf.it; salvatore.orlando@inaf.it

R. Bonito · M. Miceli
INAF-Osservatorio Astronomico di Palermo, Palermo, Italy

Dipartimento di Fisica e Chimica, Università di Palermo, Palermo, Italy
e-mail: rosaria.bonito@inaf.it; marco.miceli@inaf.it

A. I. Gómez de Castro
AEGORA Research Group, Universidad Complutense de Madrid, Madrid, Spain
e-mail: anai_gomez@mat.ucm.es

C. Sauty (ed.), *JET Simulations, Experiments, and Theory*,
Astrophysics and Space Science Proceedings 55,
https://doi.org/10.1007/978-3-030-14128-8_8

2 Numerical Modeling

Hydrodynamic (HD) models of both continuous [5, 6] and pulsed [2, 4] jets predicted X-ray emission from mechanical heating due to shocks produced by the interaction between the jet and the ambient medium. They reproduced well part of the phenomenology observed without considering magnetic field effects and they found that most of the X-ray emission is located at the base of the jet where the plasma blob collisions are the most energetic. In [3] they were able to explain the stationary component detected in HH 154 describing a shock diamond formed at the opening of a nozzle, which produces a X-ray emitting stationary source.

In [18] we proposed a new magnetohydrodynamic (MHD) model describing continuously driven jets ramming into a magnetized medium taking into account, for the first time, both magnetic-field oriented thermal conduction (including the effects of heat flux saturation) and radiative losses effects. We performed a set of 2.5D MHD numerical simulations, exploring a wide space of parameters. Later we extended the model to pulsed jets (see [17]) considering two types of numerical setups: the case of a jet less dense than the ambient medium (light jet), and the case of a jet denser than the ambient (heavy jet). The calculations were performed using PLUTO [12], a modular Godunov-type code for astrophysical plasmas. For a detailed description of the model and the numerical setup see [17, 18].

3 Main Results

We followed jet evolution for at least $\sim$50 years (see [17] for a movie). The jet propagates through the magnetized domain and expands because its dynamic pressure is much larger than the ambient pressure. During the expansion the jet density and temperature decrease, and so on the pressure, and the jet is gradually collimated by the ambient magnetic field forming a shock diamond (see [18]). The formed shock is stationary and emits in X-rays with a luminosity of $L_X \approx 10^{29}$ erg s^{-1}.

In order to study the effect of perturbations on the shock described above, we introduced a pulsed component formed from a train of blobs (for more details, see [17]). We found that in most of the cases explored the shock stability is slightly affected and that the X-ray emitting source is still present with luminosity and count rate values comparable with observations [7, 10]. In Figs. 1 and 2 we show the jet perturbed by a train of blobs for the light and heavy jet reference cases, respectively, described in detail in [17]. We observe a shock with temperatures of $T \approx 10^6$ K, emitting in X-rays. In both cases we find similar mechanisms and morphology.

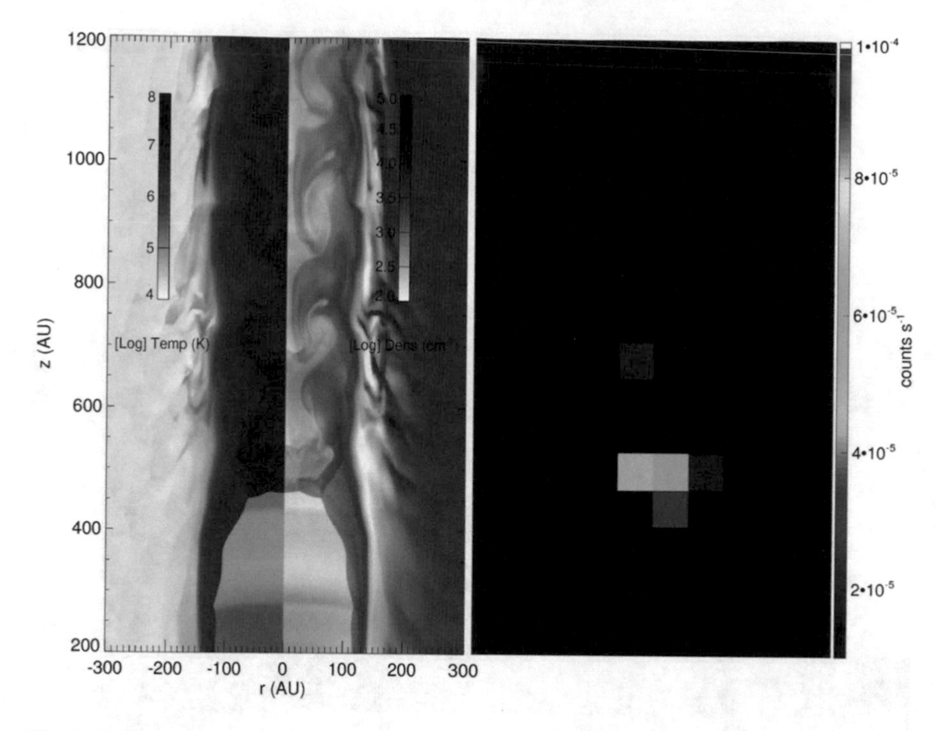

Fig. 1 Reference case for the light jet at t $\approx$ 120 yr. (Adapted from [17]). Left panel: Two-dimensional maps of temperature (left half-panel), and density (right half-panel) distributions. Right panel: Map of X-ray count rate in the [0.3–4] keV band with macropixel resolution of 0.5″ (*Chandra* typical resolution)

3.1 Comparison of the Models with Observations

We synthesized the emission measure, X-ray luminosity and count rate in all the cases, and compared model results with observations (see [17, 18] for a complete description). In particular, we studied in detail the case of HH 154 [7], a jet originating from the embedded binary Class 0/I protostar IRS 5 (light jet), and the case of the jet associated with DG Tau [11], a more evolved Class II disk-bearing source (heavy jet).

In Fig. 3 we show the count rate of the X-ray source associated with HH 154 in the [0.3–4] keV band. We compare the data set observed by *Chandra*/ACIS in 2005 [3], with the synthetic map of count rate as would be seen by *Chandra* derived from the reference case for the light jet presented in [17]. We observe both, the stationary and the variable, X-ray emitting jet components identified before by [3].

In Fig. 4 we show the count rate of the X-ray source associated with DG Tau in the [0.5–1] keV band. In the left panel we plot the merged data set observed by *Chandra*/ACIS in 2010 and analysed by [17]. For the sake of comparison, on the right we show the synthetic map derived from the reference case for the heavy jet

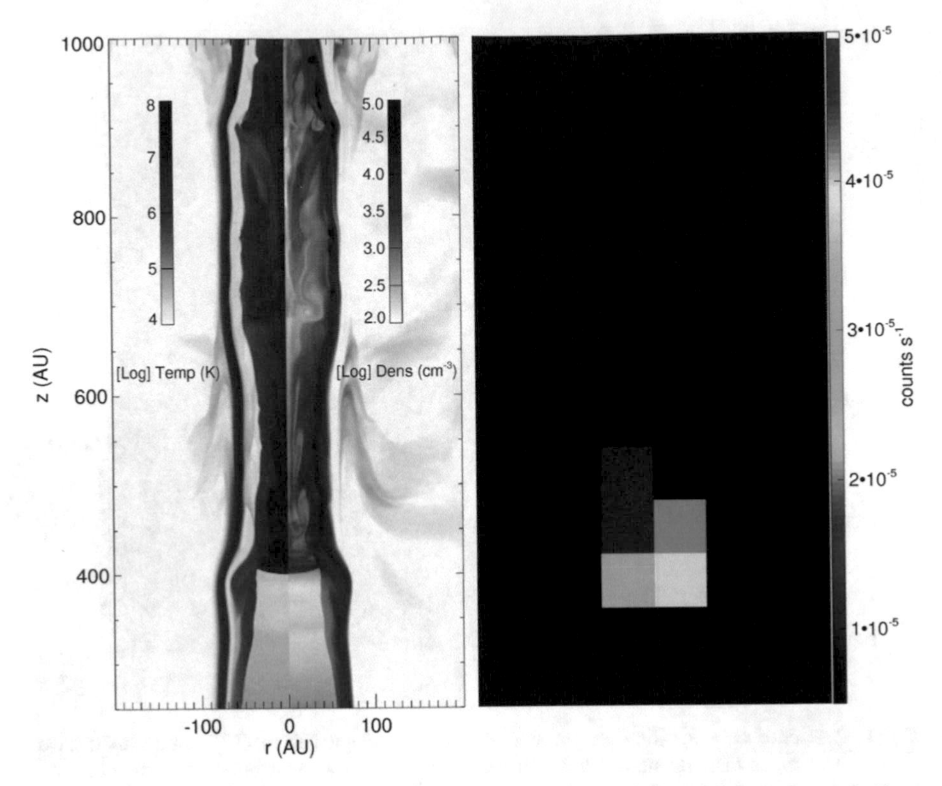

Fig. 2 Reference case for the heavy jet at t ≈ 90 yr. (Adapted from [17]). Left panel: Two-dimensional maps of temperature (left half-panel), and density (right half-panel) distributions. Right panel: Map of X-ray count rate in the [0.5–1] keV band with macropixel resolution of 0.5″ (*Chandra* typical resolution)

presented in [17] with the SW jet of DG Tau identified with a box and observed earlier by [10, 11].

4 Conclusions

We investigated the X-ray stationary sources detected at the base of some jets through MHD numerical simulations, and we compared model results with observations. The jet is collimated by the magnetic field forming a quasi-stationary X-ray emitting shock at the base of the jet which, under certain conditions, continues emitting in X-rays even when perturbations are present. The variability of the synthesized count rate is compatible with the observations of HH 154 [3] and DG Tau [9]. We found similar collimation (magnetic) mechanisms in both cases.

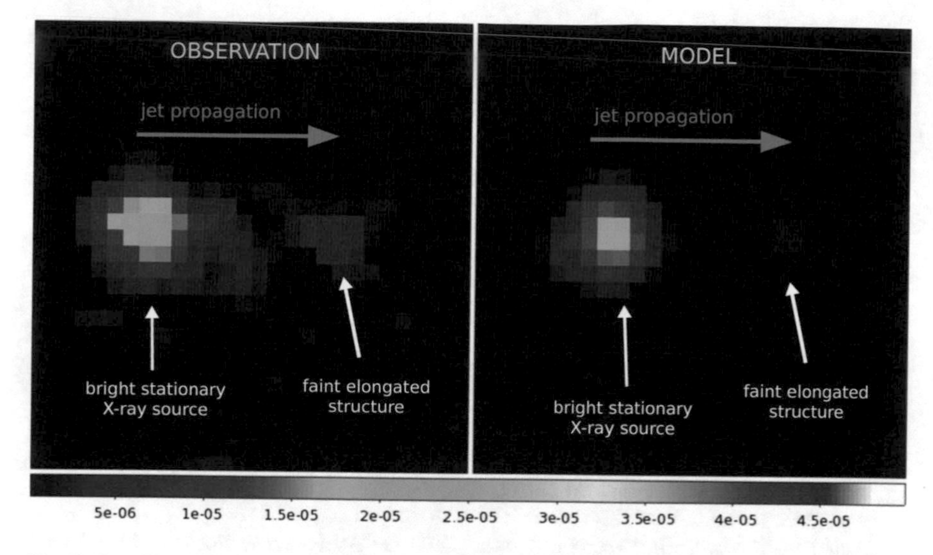

Fig. 3 Smoothed X-ray count rate maps in the [0.3–4] keV band for HH 154 with a pixel size of 0.25″. (Adapted from [17]). Left: 2005 data set resampled using the EDSER technique. Right: Synthetic image of the base of the jet derived from the model at t ≈ 100 yr, rebinned and convolved with the proper PSF. The angular size of each panel is ≈7 × 7″

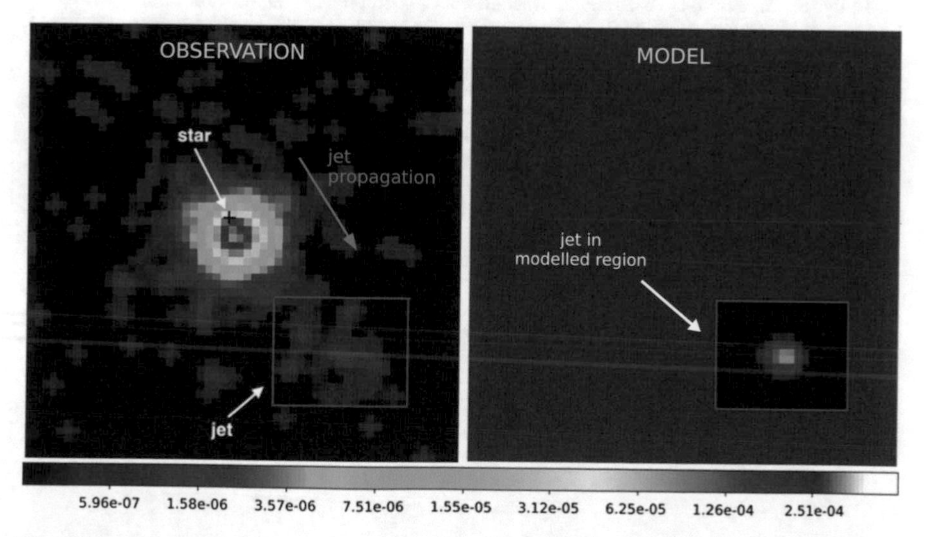

Fig. 4 Smoothed X-ray count rate maps in the [0.5–1] keV band for DG Tau in logarithmic scale and with a pixel size of 0.25″. (Adapted from [17]). Left: 2010 data set resampled using the EDSER technique. Right: Synthetic image of the base of the jet derived from the model at t ≈ 70 yr, rebinned and convolved with the proper PSF. The modelled region is marked with a box. The angular size of each panel is ≈14 × 14″

Acknowledgements PLUTO is being developed at the Turin Astronomical Observatory in collaboration with the Department of General Physics of the Turin University. This work was supported by the Ministry of Economy and Competitivity of Spain under grant numbers ESP2014-54243-R and ESP2015-68908-R.

References

1. Bally, J., Feigelson, E., & Reipurth, B. 2003, Astrophys. J., 584, 843
2. Bonito, R., Orlando, S., Miceli, M., et al. 2010a, Astron. Astrophys., 517, A68
3. Bonito, R., Orlando, S., Miceli, M., et al. 2011, Astrophys. J., 737, 54
4. Bonito, R., Orlando, S., Peres, G., et al. 2010b, Astron. Astrophys., 511, A42
5. Bonito, R., Orlando, S., Peres, G., Favata, F., & Rosner, R. 2004, Astron. Astrophys., 424, L1
6. Bonito, R., Orlando, S., Peres, G., Favata, F., & Rosner, R. 2007, Astron. Astrophys., 462, 645
7. Favata, F., Bonito, R., Micela, G., et al. 2006, Astron. Astrophys., 450, L17
8. Favata, F., Fridlund, C. V. M., Micela, G., Sciortino, S., & Kaas, A. A. 2002, Astron. Astrophys., 386, 204
9. Güdel, M., Audard, M., Bacciotti, F., et al. 2011, in Astronomical Society of the Pacific Conference Series, Vol. 448, 16th Cambridge Workshop on Cool Stars, Stellar Systems, and the Sun, ed. C. Johns-Krull, M. K. Browning, & A. A. West, 617
10. Güdel, M., Skinner, S. L., Audard, M., Briggs, K. R., & Cabrit, S. 2008, Astron. Astrophys., 478, 797
11. Güdel, M., Skinner, S. L., Briggs, K. R., et al. 2005, Astrophys. J. Letters, 626, L53
12. Mignone, A., Bodo, G., Massaglia, S., et al. 2007, Astrophys. J. Supp., 170, 228
13. Orlando, S. & Favata, F. 2008, in Lecture Notes in Physics, Berlin Springer Verlag, Vol. 742, Jets from Young Stars II, ed. F. Bacciotti, L. Testi, & E. Whelan, 173
14. Pravdo, S. H., Feigelson, E. D., Garmire, G., et al. 2001, Nature, 413, 708
15. Schneider, P. C., Günther, H. M., & Schmitt, J. H. M. M. 2011, Astron. Astrophys., 530, A123
16. Stelzer, B., Hubrig, S., Orlando, S., et al. 2009, Astron. Astrophys., 499, 529
17. Ustamujic, S., Orlando, S., Bonito, R., Miceli, M., & Gómez de Castro, A. I. 2018, Astron. Astrophys., 615, A124
18. Ustamujic, S., Orlando, S., Bonito, R., et al. 2016, Astron. Astrophys., 596, A99

Simulating Accretion and Outflow Regions in YSOs

R. M. G. de Albuquerque, V. Cayatte, J. F. Gameiro, J. J. G. Lima, C. Sauty, and S. Ulmer-Moll

1 Introduction

CTTS are low-mass pre-main sequence stars going through an active evolutionary stage. They not only accrete material from their circumstellar disks, but they also eject part of it in different shapes of outflows. The variability of YSOs along time makes the study of these stars even more complex. Therefore, both observational and theoretical sides are needed to understand most of the physical processes enrolled. In this work, we will show how numerical simulations based on an analytical model can be constrained with observations and what conclusions can be drawn from them.

R. M. G. de Albuquerque (✉)
Instituto de Astrofísica e Ciências do Espaço, Universidade do Porto, CAUP, Porto, Portugal

Departamento de Física e Astronomia, Faculdade de Ciências, Universidade do Porto, Porto, Portugal

Laboratoire Univers et Théories, Observatoire de Paris, UMR 8102 du CNRS, Université Paris Diderot, Meudon, France
e-mail: Raquel.Albuquerque@astro.up.pt

V. Cayatte · C. Sauty
Laboratoire Univers et Théories, Observatoire de Paris, UMR 8102 du CNRS, Université Paris Diderot, Meudon, France
e-mail: Veronique.Cayatte@obspm.fr; Christophe.Sauty@obspm.fr

J. F. Gameiro · J. J. G. Lima · S. Ulmer-Moll
Instituto de Astrofísica e Ciências do Espaço, Universidade do Porto, CAUP, Porto, Portugal

Faculdade de Ciências, Departamento de Física e Astronomia, Universidade do Porto, Porto, Portugal
e-mail: jgameiro@astro.up.pt; jlima@astro.up.pt; Solene.Ulmer-Moll@astro.up.pt

C. Sauty (ed.), *JET Simulations, Experiments, and Theory*, Astrophysics and Space Science Proceedings 55, https://doi.org/10.1007/978-3-030-14128-8_9

2 Deriving Stellar Activity Parameters

In this study, we will use a spectrum from RY Tau taken at the William Herschel Telescope, La Palma, in 7 November of 1998. The used slit was 6" long and 1.2" wide, returning a spectral resolution near 51,000. This observation was originally wavelength corrected for heliocentric velocity and normalized manually to a continuum level of unity. Additionally, the telluric absorptions were removed with *Molecfit* [8, 18]. From this spectrum, we derived stellar activity parameters from permitted and forbidden emission lines. These parameters include mass accretion rate, mass loss rate and projected terminal velocity of the jet.

2.1 Measuring Accretion

Probably one of the most studied accretion tracers is the emission of Hα (6563 Å). This Balmer line is characterized in RY Tau spectrum by the presence of a blue-shifted absorption and broad wings, extending until a few hundreds of km s^{-1}.

The mass accretion rate ($\dot{M}_{\rm acc}$) was measured using an empirical relation based on the width of Hα at 10% of peak intensity [13]. Through this method, we retrieved for RY Tau a value of $10^{-7.1\pm0.7}\,M_\odot\,{\rm yr}^{-1}$. This result is in agreement with the literature, where the corresponding values range between $10^{-7.7}$ and $10^{-7.0}\,M_\odot\,{\rm yr}^{-1}$ for this star [2, 4, 7, 11].

2.2 Outflow Dynamics

When dealing with outflow processes, the forbidden emission line of [OI] (6300 Å) is a good tracer to study jets and disk winds. This is possible by decomposing the profile in two emission components: the low velocity component (LVC), likely associated with the disk wind, and the high velocity component (HVC) linked with the stellar jet [6, 7, 10]. It is possible to measure the projected terminal velocity ($V_{\rm term}$) of the jet with the HVC of this forbidden emission. In order to do it, we take the value of the blue wing of the profile that crosses the continuum. We measured for RY Tau a value of $143 \pm 10\,{\rm km\,s^{-1}}$. This value fits reasonably the velocity of the blue-shifted jet of RY Tau, measured recently from the C IV line, with a value of $-136 \pm 10\,{\rm km\,s^{-1}}$ at 39 AU [17].

To derive the mass loss rate, we need to measure the equivalent width of [OI] at 6300 Å ($EW_{\rm [OI]} = 0.7$ Å), convert it to line luminosity (L_{6300}) and then compute the mass loss rate ($\dot{M}_{\rm loss}$), following relations already available [3, 7]. The adopted equation with the inclusion of a veiling correction for the luminosity of the forbidden line is given by

$$L_{6300}(L_\odot) = 6.71 \times 10^{-5} D^2 EW_{\rm [OI]}(1 + r_{6200})10^{-0.4R_0}, \quad (1)$$

where D is the distance star-observer of 140 pc [9], R_0 is the derredened magnitude of 9.83 [7] and r_{6200} is the veiling of 0.21 we estimated near 6200 Å with a photospheric template. Once the luminosity is determined, we derive the mass loss rates following the equation below

$$\dot{M}_{\rm loss}(M_{\odot}\,{\rm yr}^{-1}) = 3.03 \times 10^{7}\left(1 + \frac{N_c}{N_e}\right) \times \frac{L_{6300}(L_{\odot})V(\rm km\,s^{-1})}{l(\rm cm)}, \qquad (2)$$

where $N_c = 2 \times 10^6\,\rm cm^{-3}$ is the critical density for [OI] [3] and $N_e = 7 \times 10^4\,\rm cm^{-3}$ is the electronic density in the emitting volume [7]. V and l are the component of the outflow velocity and the projected size of the aperture on the plane of the sky, respectively. In this work, we project the measured terminal velocity on the plane of the sky ($V = V_{\rm term} \tan i$), where i is the inclination of the stellar jet, and use it as input in Eq. (2). We will assume the inclination determined for the blue-shifted jet of RY Tau of $i = 61° \pm 16°$ [17]. The last parameter is given by $l = 1.2'' D\, 1.5 \times 10^{13}$ cm. Finally, we derive a mass loss of $10^{-7.8}\, M_{\odot}\, yr^{-1}$. Taking into account the mass accretion rate estimated previously, we derive a $\dot{M}_{\rm loss}/\dot{M}_{\rm acc}$ rate of 0.2. The value we obtain for mass loss rate is quite high and likely associated with a large error. This could be related to the fact that we use a low electronic density or underestimated projected size of the aperture.

3 Constraining Numerical Simulations

For the modeling of accretion and outflow regions in the surroundings of a CTTS, we used a meridional self-similar model [16] coupled with a stable analytical solution developed for CTTS [15]. The outflows in the model are pressure-driven and the analytical solution considers some observational parameters of RY Tau, namely mass loss rate, stellar radius and mass of $10^{-8.5}\, M_{\odot}\,\rm yr^{-1}$ [5], 2.4 $R_{\odot}$ [7] and 1.5 $M_{\odot}$, respectively.

The 2.5 MHD numerical simulations were carried out in PLUTO code [12]. Firstly, accretion was implemented in the simulations by reversing the sign of the poloidal velocity, inside the closed magnetospheric region. Secondly, the multiplying factors of density and poloidal velocity were increased and adjusted such that the accretion rates and terminal velocities of the jet measured match with the observational ones. In Table 1 are listed four simulations. Each one has a different combination of multiplying factors for the poloidal velocity (V_p) and density (ρ). The corresponding mass loss ($\dot{M}_{\rm loss}$), accretion rates ($\dot{M}_{\rm acc}$) and the ratio between these two quantities are listed as well.

We can see that, as we increase the poloidal velocity and density multiplying factors, we increase the accretion rates and the corresponding rates between mass loss and accretion decrease. The rates of Simulations C and D are in agreement with the literature for RY Tau, between 0.02 and 0.4 [1]. From the observations we

Table 1 Mass fluxes measured for each simulation

Simulation	V_p	ρ	$\dot{M}_{\rm loss}(M_\odot\,{\rm yr}^{-1})$	$\dot{M}_{\rm acc}$	$\dot{M}_{\rm loss}/\dot{M}_{\rm acc}(M_\odot\,{\rm yr}^{-1})$
A	1.0	1	$10^{-8.6}$	$10^{-8.4}$	0.71
B	1.5	1	$10^{-8.6}$	$10^{-8.3}$	0.45
C	1.5	5	$10^{-8.6}$	$10^{-7.7}$	0.12
D	2.0	10	$10^{-8.6}$	$10^{-7.3}$	0.05

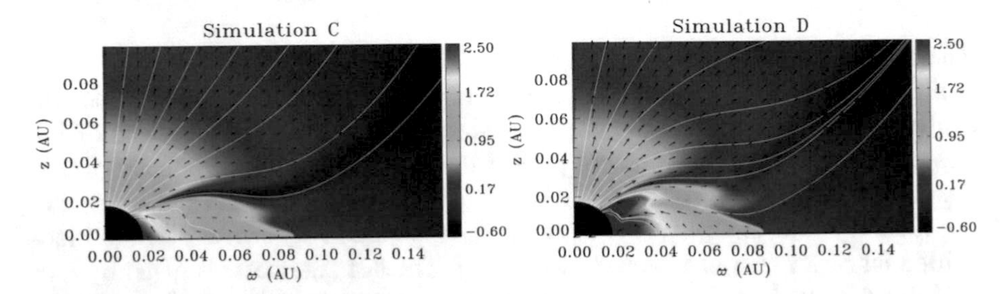

Fig. 1 Logarithmic density maps for simulations C and D

calculated a ratio of 0.2, which fits in the previous interval. The mass loss remains nearly constant along time due to the preservation of the stellar jet region. We observed that the latest is not too much affected by these accretion increments. Another particular aspect involves the combination of the poloidal velocity and density factors. According to the simulations, they seem to associate in a specific way, such that if we combine them differently they can return a more unstable configuration.

In Fig. 1 are shown the logarithmic density plots, in PLUTO units, for simulations C and D at time 55, which corresponds to 7 stellar rotations, approximately. The white solid lines correspond to the magnetic field lines, while the black arrows represent the velocity vectors. The horizontal and vertical axis are in astronomical units. Simulation C reached a steady configuration after 2.5 stellar rotations. Interestingly, there is a low density corridor that separates the stellar jet region from the closed magnetospheric region. Conversely, Simulation D shows a more perturbed configuration with sporadic magnetospheric ejections being released towards the interstellar medium. At the stellar jet region, we measured for these two simulations the projected terminal velocity of the jet at 36 AU. We retrieved $186^{+86}_{-99}\,{\rm km\,s^{-1}}$ for simulation C, and $181^{+83}_{-97}\,{\rm km\,s^{-1}}$ for simulation D. Both simulations are in agreement with our observational measurement, although there are large error bars. This is due to the poor accuracy on the jet inclination which does not allow stronger constraints on the current simulations.

4 Conclusions

In this study for RY Tau, we were able to simulate not only accretion, but also outflows through PLUTO code and a self-similar model. To achieve a stable solution, it seems that there is a correspondence between velocity and density factors. Additionally, we conclude that simulations C and D are the best results reached so far. The rates between ejected and accreted material fit within the intervals available in the literature, as well as the mass accretion rates. On one hand, simulation C shows a steady configuration with a low density corridor. On the other hand, simulation D is characterized by the release of magnetospheric ejections. Taking all into account, these two simulations may support the idea that RY Tau has a bimodal behaviour [14], oscillating between a quiescent epoch and a more active one with enhanced outflowing activity.

Acknowledgements This work was supported by FCT – Fundação para a Ciência e a Tecnologia through national funds and by FEDER – Fundo Europeu de Desenvolvimento Regional through COMPETE2020 – Programa Operacional Competitividade e Internacionalização by these grants: UID/FIS/04434/2019 and PTDC/FIS-AST/32113/2017 & POCI-01-0145-FEDER-028987. RMGA is supported by the fellowship PD/BD/113745/2015, under the FCT PD Program PhD::SPACE (PD/00040/2012), funded by FCT (Portugal) and POPH/FSE (EC). We acknowledge financial support from Programme National de Physique Stellaire (PNPS) of CNRS/INSU (France) and from CRUP through the PAUILF cooperation program (TC-16/17). Some of the computations made use of the High Performance Computing OCCIGEN at CINES within the Dari project c2016047602. The authors would like to thank A. Pedrosa for providing the observational data.

References

1. Agra-Amboage, V., Dougados, C., Cabrit, S., Garcia, P. J. V., & Ferruit, P. 2009, A&A, 493, 1029
2. Calvet, N., Muzerolle, J., Briceño, C., et al. 2004, AJ, 128, 1294
3. Comerón, F., Fernández, M., Baraffe, I., Neuhäuser, R., & Kaas, A. A. 2003, A&A, 406, 1001
4. Costigan, G., Vink, J. S., Scholz, A., Ray, T., & Testi, L. 2014, MNRAS, 440, 3444
5. Gómez de Castro, A. I. & Verdugo, E. 2001, ApJ, 548, 976
6. Hamann, F. 1994, ApJs, 93, 485
7. Hartigan, P., Edwards, S., & Ghandour, L. 1995, ApJ, 452, 736
8. Kausch, W., Noll, S., Smette, A., et al. 2015, A&A, 576, A78
9. Kenyon, S. J., Dobrzycka, D., & Hartmann, L. 1994, AJ, 108, 1872
10. Kwan, J. & Tademaru, E. 1988, ApJl, 332, L41
11. Mendigutía, I., Eiroa, C., Montesinos, B., et al. 2011, A&A, 529, A34
12. Mignone, A., Bodo, G., Massaglia, S., et al. 2007, ApJs, 170, 228
13. Natta, A., Testi, L., Muzerolle, J., et al. 2004, A&A, 424, 603
14. Petrov, P. P., Grankin, K. N., Gameiro, J. F., et al. 2019, MNRAS, 483, 132
15. Sauty, C., Meliani, Z., Lima, J. J. G., et al. 2011, A&A, 533, A46
16. Sauty, C. & Tsinganos, K. 1994, A&A, 287, 893
17. Skinner, S. L., Schneider, P. C., Audard, M., & Güdel, M. 2018, ApJ, 855, 143
18. Smette, A., Sana, H., Noll, S., et al. 2015, A&A, 576, A77

Modeling Jet Launching from Accretion Disks

C. Fendt

1 Jet Launching: From Accretion to Ejection

Jets are powerful signatures of astrophysical activity and can be observed over a wide range of energy output and spatial extent. Jets from young stellar objects are particularly interesting, as they allow us to observe a range of dynamical properties that are essential for modeling. It is commonly accepted that MHD processes are essential for launching, accelerating and collimating outflows from accretion disks [2, 7, 9, 17–19, 27].

We suggest to distinguish between *jet launching* – the transition from accretion to ejection – and *jet formation* standing for the acceleration and collimation process of a disk wind into a narrow jet. Jet *formation* is usually understood as due to the magneto-centrifugal slingshot mechanism [2, 17]. Outflows of lower velocity are built up as so-called tower jets accelerated by the magnetic pressure gradient of the toroidal field [13].

Jet *launching* is more complex as interrelating the disk physics with the jet physics. For example, disks are viscous and magnetically diffusive, while jets and outflows can be treated in ideal MHD. Also, disks may have a turbulent, tangled field component, while fast jets rely on the existence of a large-scale magnetic field. The magnetic diffusivity is believed to be of turbulent origin. Jet launching is not yet fully understood in detail, however, (semi) analytical modeling [7, 12] and numerical simulations [3, 11, 30] agree on the main mechanisms involved.

The general idea is that gas is advected along the disk plane. A large-scale magnetic field penetrates the disk. The field distribution is affected by advection and diffusion. Essentially, the lifting of disk material into the outflow is a magnetic

C. Fendt (✉)
MPI for Astronomy, Heidelberg, Germany
e-mail: fendt@mpia.de

C. Sauty (ed.), *JET Simulations, Experiments, and Theory*,
Astrophysics and Space Science Proceedings 55,
https://doi.org/10.1007/978-3-030-14128-8_10

process for which Lorentz forces play a leading role [7, 8]. Once lifted above the disk surface, acceleration is typically by the magneto-centrifugal effect. The collimation into a narrow beam is accomplished by the pinching forces of the toroidal magnetic field. Simulations of the acceleration and collimation processes of jets forming from the disk surface have been performed by a number of authors (see e.g. [4, 5, 10, 16, 23, 28]).

In the next sections we present example results of our recent jet launching simulations. Figure 1 shows the evolution of the inner part of a jet launching disk. The magnetic field first diffuses outwards (shown are flux surfaces $\Psi \propto \int B_z dr$), until a disk wind starts removing angular momentum and accretion begins, consequently leading to the advection of magnetic flux. Essentially, the outflow *launching conditions change over time* until a steady-state is reached.

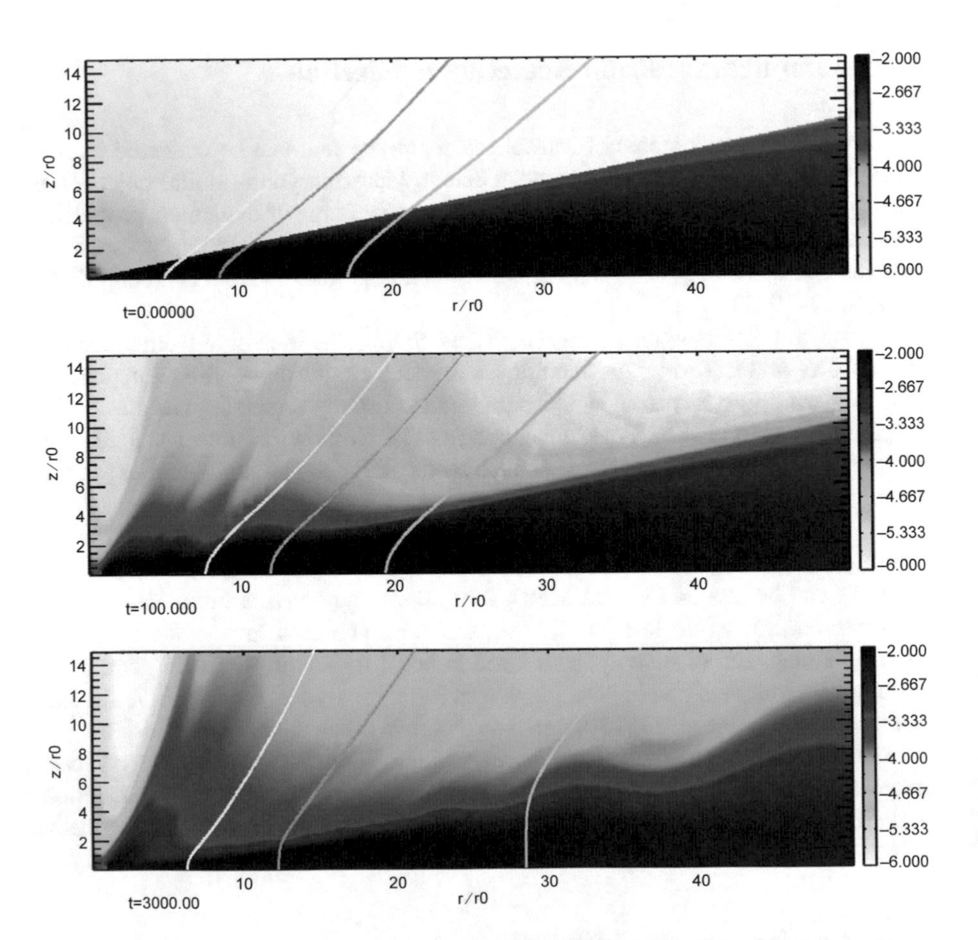

Fig. 1 Advection and diffusion of outflow-launching magnetic flux surfaces. Three flux surfaces are shown (disk density in color). The panels show the inner part (length scale normalized to inner disk radii) of a simulation reaching $r = 100$, $z = 300$ in extend [20]

2 Jet Launching and Disk Magnetization

We have invented a novel approach to the jet-launching problem in order to obtain correlations between the physical properties of the jet and the underlying disk [25]. We have investigated a wide parameter range of the *disk magnetization* $\mu_{\rm D}$ at the outflow launching radius. The magnetization μ measures the ratio of magnetic pressure to gas pressure. Our study is complementary to the works of [26] and [15] who investigated the launching conditions along a disk surface.

We have disentangled the disk magnetization as the main parameter governing the outflow properties. Strongly magnetized disks launch more energetic and faster jets, and, due to a larger Alfvén lever arm, these jets extract more angular momentum from the disk. These kinds of systems have, however, a weaker mass loading and a lower mass ejection–accretion ratio. Jets are launched at the disk surface where the magnetization is $\mu(r, z) \simeq 0.1$.

Figure 2 shows the tight correlations between the total jet energy e and the mass loading k with the disk magnetization $\mu_{\rm D}$, respectively. The mass loading parameter k measures the amount of matter ejected per unit magnetic flux. Each of the differently colored paths corresponds to a simulation with a different initial disk magnetization. We also find indication of a critical disk magnetization $\mu_D \simeq 0.01$ separating the regimes of magneto-centrifugally driven and magnetic pressure-driven jets (indicated by the change in slope in Fig. 2). The existence of these two regimes has been discussed by Ferreira [7], however, we obtain these correlations from simulations that include the dynamical evolution of disk and jet.

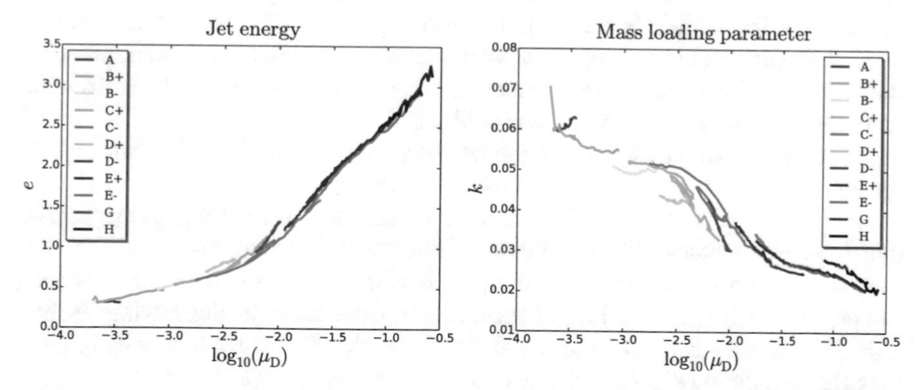

Fig. 2 Jet properties with respect to the disk magnetization $\mu_{\rm D}$. The jet energy e, and the mass loading parameter k are shown. Each colored line represents the evolution of a single simulation up to 10,000 time units. (Taken from [25])

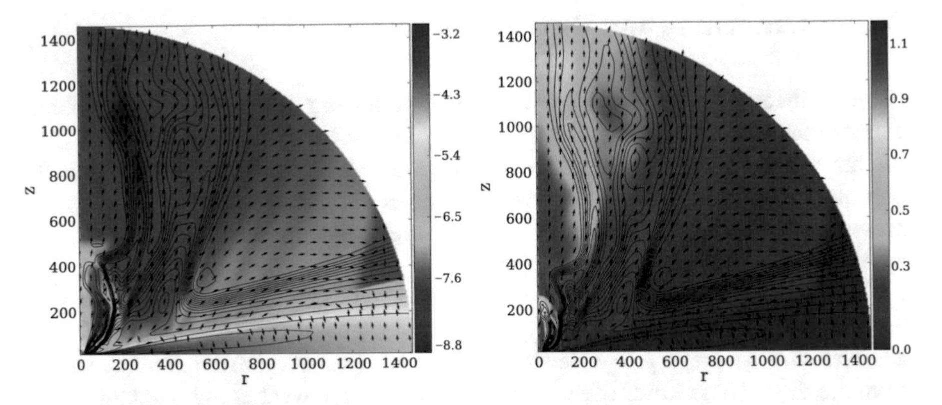

Fig. 3 Jet launching from dynamo-active disks. Shown is the density (left) and outflow velocity (right) in code units (colors). The velocity is normalized to the Keplerian velocity at the inner disk radius, The dynamo model is time dependent with a period of about 100 orbital times. Superimposed are poloidal magnetic field lines (black), see also Stepanovs et al. [24]

3 Launching by a Disk-Dynamo Generated Magnetic Field

Most simulations of jet launching have applied a prescribed large-scale initial magnetic flux. Exceptions are e.g. [1] or [29] considering a disk dynamo that generate the jet driving magnetic field. Recently we have presented MHD simulations exploring the launching, of jets considering the generation of the magnetic field by an α^2-Ω mean-field disk dynamo [24]. We find a dynamo-generated field structure of the inner disk similar to the commonly prescribed open field structure, favoring magneto-centrifugal launching. The outer disk field is inclined and predominantly radial. Here, differential rotation induces a strong toroidal field. Outflows from the outer disk are slower, denser, and less collimated.

We have further applied a toy model triggering a time-dependent mean-field dynamo. When the dynamo is suppressed, the magnetization falls below a critical value, and the generation of the outflows is inhibited. Dynamo-activity follows dynamo-inactive times with periodicities of about 200 disk rotations (Fig. 3). During dynamo-active periods new jet ejections happen. In the figure two ejections periods can be identified (more in velocity than in density), the latest ejection having reached $z = 200\,R_{\text{in}}$, the earlier one located at $z = 1200\,R_{\text{in}}$. This work has recently been updated considering bipolar dynamos and outflows [6].

4 3D-Launching Simulations

Extending the launching setup to 3D, we have recently presented the first ever 3D simulations of the MHD accretion-ejection structure [21]. We have implemented a 3D gravitational potential due to a companion star and run simulations with different

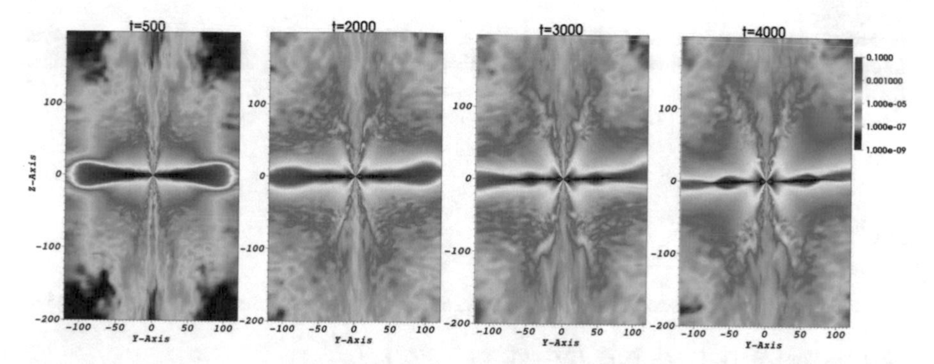

Fig. 4 3D jet launching from jet sources in binary systems. Shown is a time sequence of slices through the 3D simulation for the density structure. (Taken from [21])

binary separation. We see typical 3D deviations from axial symmetry, such as jet bending outside the Roche lobe or spiral arms forming in the accretion disk. An exemplary parameter setup with a small binary separation of only $\simeq$200 inner disk radii indicates the onset of jet precession – caused by the wobbling of the jet-launching disk. A final prove for precession can only be given by much longer simulations lasting several orbital time scales. In Fig. 4 we see the disk-realignment over time from the initial orientation towards the orbital plane. The secondary is located at $x = 200$ and $z = 60$ (outside the numerical grid).

Our simulations suggest the existence of a critical inclination angle between the disk and the binary orbit beyond which tidal forces disturb jet formation [22].

5 Outlook

We have discussed *global* models of jets launching in which the "microphysics" of the disk is approximated by averaged quantities such as (mean) magnetic flux, plasma density, flow energy or angular momentum, and, in particular, mean turbulent diffusivity or a mean-field turbulent dynamo. This seems feasible and has so far provided promising results that are in nice agreement with analytical theory.

Certain physical aspects that may play role in reality have not yet been addressed in detail in global numerical launching models. Some are, however, currently investigated in (local and global) accretion disk simulations and will certainly be the topic of future disk-jet launching modeling. Examples are e.g. (i) heating and cooling of the disk, (ii) a self-consistent description of the turbulent magnetic diffusivity, and similarly (iii) for the mean-field turbulent dynamo, then (iv) non-ideal MHD effects as ambipolar diffusion or Hall MHD, the influence of (v) the stellar magnetic field, or (vi) the influence of radiation pressure.

A full understanding of jet launching will require a more complete treatment of the internal disk physics.

Acknowledgements The work discussed above has mostly be done by my students Somayeh Sheiknezami, Bhargav Vaidya, Oliver Porth, Deniss Stepanovs, and Dennis Gaßmann. I thank them for a fruitful collaboration. For the numerical work we applied the PLUTO code [14]. I thank Andrea Mignone and the PLUTO team for the possibility of using their code.

References

1. Bardou, A.,von Rekowski, B.,Dobler, W., Brandenburg, A. & Shukurov, A. 2001, A&A, 370, 635
2. Blandford, R. D., Payne, D. G. 1982, MNRAS, 199, 883
3. Casse, F. & Keppens, R. 2002, ApJ, 581, 988
4. Fendt, C., Elstner, D. 1999, A&A, 349, L61
5. Fendt, C., Čemeljić, M., 2002, A&A, 395, 1045
6. Fendt, C., Gaßmann, D. 2018, ApJ, 855, 130
7. Ferreira, J. 1997, A&A, 319, 340
8. Ferreira, J., Dougados, C., Cabrit, S. 2006, A&A, 453, 785
9. Hawley, J. F., Fendt, C., Hardcastle, M., Nokhrina, E. & Tchekhovskoy, A. 2015, SSRv, 191, 441
10. Krasnopolsky, R., Li, Z.-Y., Blandford, R. 1999, ApJ, 526, 631
11. Kuwabara, T., Shibata, K., Kudoh, T., Matsumoto, R. 2005, ApJ, 621, 921
12. Li, Z.-Y. 1995, ApJ, 444, 848
13. Lynden-Bell, D. 1996, MNRAS, 279, 389
14. Mignone, A., Bodo, G., Massaglia, S., Matsakos, T., Tesileanu, O., Zanni, C., & Ferrari, A. 2007, ApJS, 170, 228
15. Murphy, G. C., Ferreira, J., & Zanni, C. 2010, A&A, 512, A82
16. Ouyed, R., Pudritz, R. E., 1997, ApJ, 482, 712
17. Pudritz, R. E. & Norman, C. A. 1983, ApJ, 274, 677
18. Pudritz, R. E., Ouyed, R., Fendt, C., & Brandenburg, A. 2007, Protostars and Planets V, 277
19. Sauty, C., Tsinganos, K. 1994, A&A, 287, 893
20. Sheikhnezami, S., Fendt, C., Porth, O., Vaidya, B., & Ghanbari, J. 2012, ApJ, 757, 65
21. Sheikhnezami, S. & Fendt, C. 2015, ApJ, 814, 113
22. Sheikhnezami, S. & Fendt, C. 2018, ApJ, 861, 11
23. Stepanovs, D. & Fendt, C. 2014, ApJ, 793, 31
24. Stepanovs, D., Fendt, C. & Sheikhnezami, S. 2014, ApJ, 796, 29
25. Stepanovs, D. & Fendt, C. 2016, Apj, 825, 14
26. Tzeferacos, P. Ferrari, A. Mignone, A. Zanni, C. Bodo, G., Massaglia, S. 2009, MNRAS, 400, 820
27. Uchida, Y., & Shibata, K. 1985, PASJ, 37, 515
28. Ustyugova, G. V., Koldoba, A. V., Romanova, M. M., Chechetkin, V. M., Lovelace, R. V. E. 1995, ApJL, 439, L39
29. von Rekowski, B., Brandenburg, A., Dobler, W., Dobler, W. & Shukurov, A. 2003, A&A, 398, 825
30. Zanni, C., Ferrari, A., Rosner, R., Bodo, G., & Massaglia, S. 2007, A&A, 469, 811

Relativistic 3D Hydrodynamic Simulations of the W50-SS433 System

Dimitrios Millas, Oliver Porth, and Rony Keppens

1 Introduction

The W50-SS433 system is a quite peculiar astrophysical object, located in our Galaxy, consisting of the W50 *supernova remnant* (SNR) and the *X-ray binary* SS433. Observations have revealed that this object has an elongated shape, often referred to as a "manatee", where in addition a clear east-west asymmetry is present. The SS433 X-ray binary is know to host a precessing, relativistic jet [3, 9, 10], with its mean axis almost parallel to the elongated part of the SNR. The elongated shape of the remnant is often associated with the presence of the jet and its interaction with the shell and has already been examined in 2D simulations, e.g. [6]. Here we report on the first full 3D relativistic hydrodynamic simulations, aiming to capture in more detail the interaction between the SNR and the jet. Our work includes studying Rayleigh-Taylor instabilities that develop during the supernova expansion phase. The advantage of 3D runs is the possibility to inject a helical jet.

D. Millas (✉) · R. Keppens
KU Leuven, Leuven, Belgium
e-mail: dimitrios.millas@kuleuven.be; rony.keppens@kuleuven.be

O. Porth
Goethe-Universität Frankfurt am Main, Frankfurt am Main, Germany
e-mail: porth@th.physik.uni-frankfurt.de

C. Sauty (ed.), *JET Simulations, Experiments, and Theory*,
Astrophysics and Space Science Proceedings 55,
https://doi.org/10.1007/978-3-030-14128-8_11

2 Simulations

For the simulations we use the relativistic hydrodynamic module from the open source, parallel, grid adaptive, MPI-AMRVAC code [7, 11]. The computational domain is a Cartesian grid with dimensions $-100\,\text{pc} \leq x, y, z \leq 100\,\text{pc}$, where we have tested different base resolutions and adaptive mesh refinement (AMR) levels.

To examine the initial propagation of the supernova, we used a base resolution of 96^3 with 6 additional AMR levels, achieving a maximum resolution of 6144^3, resolving ~0.03 pc per cell. The additional AMR levels are activated inside a user-defined region extending up to the supernova shell.

For the full simulation (where we include the jet) we used a base resolution of 200^3 and 5 additional adaptive mesh levels (AMR), leading to an effective resolution of 3200^3, resolving ~0.06 pc per cell. This feature is necessary to capture the initial supernova blast and the injection of the jet at a later time (treated as an internal boundary), while keeping the computational cost manageable. This is partially achieved by scaling-up the injection region of the jet; instead of injecting the jet near the accretion disk, our radius of injection is set to $r \simeq 0.05\,\text{pc}$. The mean jet axis is considered to be parallel to $\hat{x}$. Our simulations handle the full precession of the jet; this was also shown earlier by [10].

For the ISM we examine two different scenarios: (i) a uniform and (ii) an exponential density profile. In both scenarios, the number density in the center of the domain is fixed to $n_o = 1$ particle/cm^3. The exponential profile for the ISM can be described as follows:

$$\rho(r, z) = \rho_o \exp(-R_m/R_d - r/R_d - z/Z_d) \quad , \tag{1}$$

where r is the distance from the galactic centre, z is the distance from the galactic disk and $R_m = 4\,\text{kpc}$, $R_d = 5.4\,\text{kpc}$, $Z_d = 40\,\text{pc}$ are constants determining the relevant scalelengths [2, 6]. An inverse behaviour is considered for the temperature profile to keep a constant pressure, so the gravitational term can be omitted. We note that the dependence on r is not expected to have a significant effect since the domain size (and by extension the size of the nebula) is small compared to the R_d scalelength.

For an equation of state (EOS), we used either (a) the Mathews [8] or (b) the ideal gas description, with no significant difference (in terms of the final shape of the shell) between the two options. In the following sections, we only present results using the Mathews EOS. Basic parameters are summarized in Table 1.

Table 1 Basic parameters of the simulations

Run	ρ profile	n_o (origin) ($\#cm^{-3}$)	E_{blast} (erg)	L^{kin}_{jet} (erg/s)	v_{SN}	v_{jet}
SNR	Exponential	1	10^{51}	–	0.015c	–
SNRJet_uni	Uniform	1	10^{51}	10^{39}	0.015c	0.25c
SNRJet_exp	Exponential	1	10^{51}	10^{39}	0.015c	0.25c

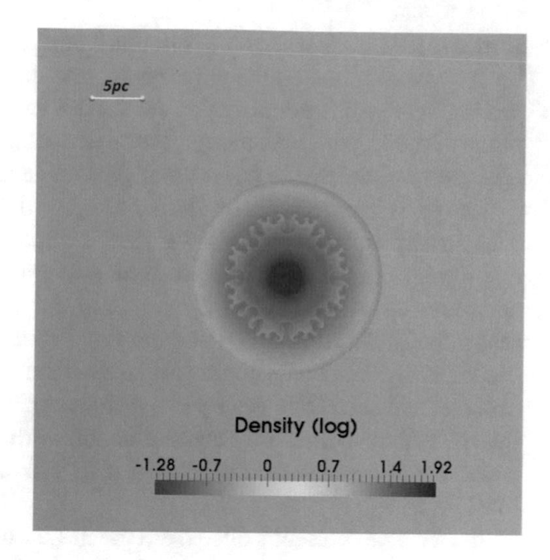

Fig. 1 Slice of a 3D run (xy plane) showing the logarithm of density at $t \sim 2500$ yrs, in a zoomed region $-25\text{pc} \leq x, y \leq 25$ pc. At this snapshot, Rayleigh-Taylor instabilities can be clearly seen. The radius of the shell at this time is $\sim$10 pc. The diagonal towards the top right corner points to the galactic disk

2.1 *The Supernova Explosion*

The supernova explosion is modelled as the expansion of a sphere of high density and velocity, inside the (static) interstellar medium. To correctly calculate the velocity of the expanding sphere, an assumption needs to be made regarding the mass of the ejected material and the energy released in the supernova blast. Since there is no direct method to constrain these quantities, we will use typical values; we assume a mass of $M_{ej} = 6M_{\odot}$ for the ejecta and 10^{51} ergs for the energy of the supernova blast. The velocity is thus of the order of $v_{SN} \sim 10^4$ km/s.

The initial radius of the sphere is taken equal to a fraction of the Sedov radius, which is calculated via the mass of the ejecta $r_{SN} = \frac{1}{4}\left(\frac{3M_{ej}}{4\pi\rho_o}\right)^{1/3}$. Using a sufficiently high resolution, we can capture Rayleigh-Taylor instabilities developing as early as $t \sim 1000$ yrs. A snapshot assuming propagation in an ISM with exponential density profile is given in Fig. 1.

Due to the asymmetric density profile of the ISM, the shell is not symmetric; the shell propagates at a slightly greater distance towards $-\hat{x}$ since the density there is lower. At $t \sim 2500$, the "east-west" relative difference in the radii is $\sim$4%; when the jet is injected, the shape of the shell can be approximated with an ellipse with eccentricity $\sim$0.25.

2.2 *The Final Shape of the Remnant*

The jet is included in the simulation as an internal boundary, activated after the supernova blast. The time of the injection is estimated by some test runs, where we measure the time needed for the SNR to reach a distance of approximately 45 pc; this is motivated by the size of the "circular" part of the shell [6]. This naturally depends on the assumptions on the supernova blast energy, the mass of the ejecta and the scaling values of the density profile; for the values given above, the injection occurs after ~20,000 yrs. Even though the shell still propagates after the jet injection, the effect is minimal due to the scale difference in the velocities.

The jet is injected with a velocity of $v_j \simeq 0.25c$ (or Lorentz factor of $\gamma \sim 1.03$). The density is calculated via the kinetic luminosity of the jet, $L_{kin} = 10^{39}$ erg/s and the injection radius. The precession of the jet is also taken into account, using as values $\theta_p = 20.92$ and $T_p = 162d$ for the precession angle and the period respectively. To reduce the computational cost, we scale up the injection region of the jet, using the kinetic luminosity to consistently calculate its density. This way the precession can be followed up to a distance of ~5 pc; although quite exaggerated, it doesn't affect significantly the interaction with the SNR since it forms a "hollow", straight outflow at these distance [9, 10]. This is also in agreement with the upper limit given by [1].

In Fig. 2 we present a slice from the 3D output, perpendicular to the $\hat{z}$ axis, with the density and the Lorentz factor for the two ISM conditions (uniform and density profile given by Eq. 1).

We verify that in order to obtain the observed asymmetry of the SNR, the density gradient is necessary. To quantify the asymmetry we calculate the difference in the distance of the edge of remnant along x; our simulations show for the "east" and "west" radii r_e, with a relative difference of ~50%. The asymmetry after ~27,000 years results is a ratio of $\simeq$1.55 for the end points of the shell along the jet axis, slightly higher than the observed value of $\simeq$1.40. This result depends on the choice of scaleheights and the value of number density n_o (where we used a higher value than the optimal choice suggested by [6]). The time of the injection of the jet is also an important parameter.

2.3 Emission

The supernova shell of the system is studied mainly in radio wavelengths, where intensity variations and depolarization in lower frequencies have been observed (see [4]).

Here we use a simple synchrotron emission recipe, assuming equipartition, to create emission maps from the simulations. We post-process the output of each run, substituting the magnetic field energy density (B) with a fraction of the internal energy density of the fluid (ϵ), $B \sim \sqrt{8\pi\epsilon}$, (see [5]). The drawback of this method is that it cannot be used to extract information on the polarization properties of the system. An emission map for the exponential case, is shown in Fig. 3 and a zoomed image of the jet is shown in Fig. 3.

The emission map shows a diffuse emission, as seen in W50 [3, 4]. We notice though some differences with the actual object: the part of the nebula facing towards the Galactic disk ($x > 0$ in our simulations) should be much brighter compared to

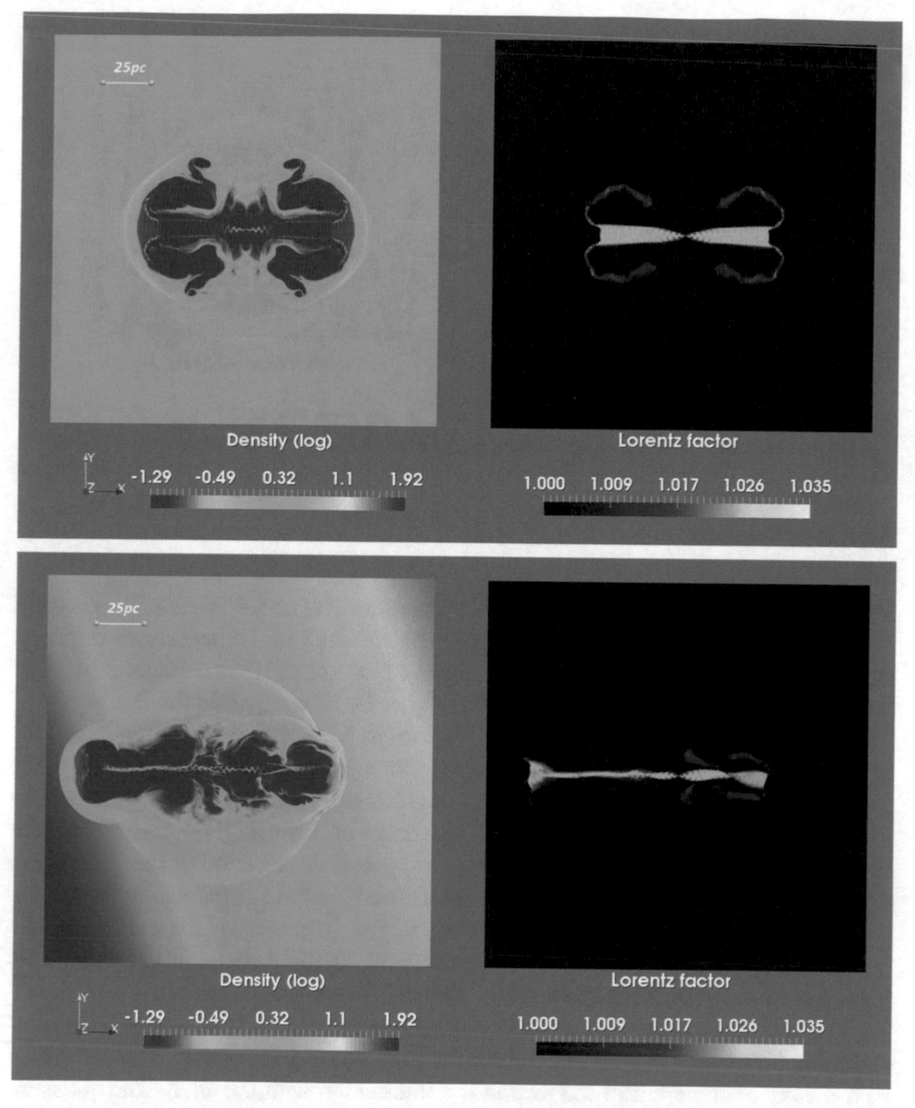

Fig. 2 Slice of a 3D run: uniform (top) vs exponential (bottom) density profile for the ISM. For each case we present the density (in log scale) and the Lorentz factor. The diagonal towards the top right corner points to the galactic disk

the opposite side. The circular part of the SNR has a very weak emission (not visible in our linear scale), which is also not the case in W50. On the other had, the bright parts of the SNR, roughly where the jet interacts with the shell, are captured. This is an issue to be examined closely in subsequent studies.

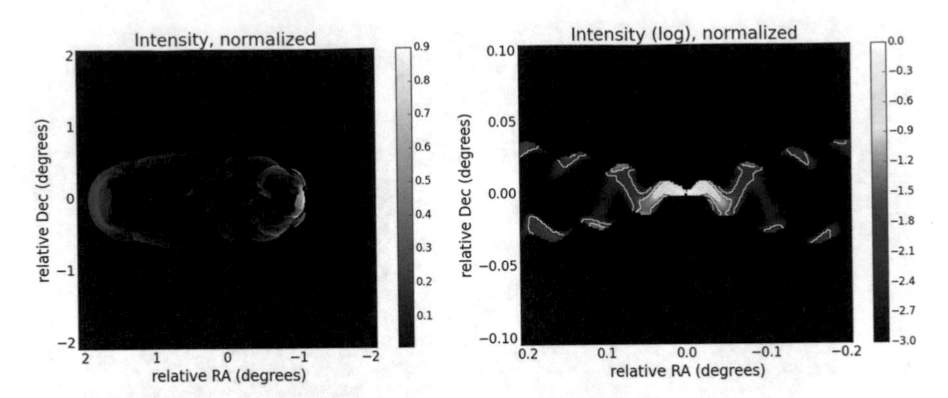

Fig. 3 Emission map of the full system in radio, at the approximate location of W50-SS433 (∼5.5 kpc) (left) and a zoomed image of the jet (right), at 20 cm and 5 GHz respectively

3 Conclusions

We performed relativistic, hydrodynamic simulations of the W50-SS433 system in 3D, aiming to resolve the interaction between the jet and the supernova remnant. Runs with higher resolution in the first stages of the simulations were used to study the evolution of Rayleigh-Taylor instabilities on the supernova shell.

We verify, in good agreement with [6], the importance of a non-uniform ISM to produce the observed asymmetric shape, using a realistic exponential profile for our Galaxy. The obtained "east-west" asymmetry from our runs, quantified by the relative difference of the radii, is of the order of 50%, slightly higher than the observed. This result can be improved by an appropriate selection of the parameters n_o, M_{ej}, the scaleheights of the density profile and the time of the injection.

Although dynamically important during the early stages of the evolution, the precession of the jet might not affect significantly the final shape of the SNR. This is related to the collimation distance of the jet, after which it becomes a "hollow" jet. This has also been examined in [1, 10].

Future work includes higher resolution runs, aiming to capture smaller scales near the injection region of the jet and the interaction with the shell. The emission from the system will be studied in more detail to create synthetic maps.

References

1. Bowler, M. G. and Keppens, R. 2018, *A&A*, 617, A29
2. Dehnen, W. and Binney, J. 1998, *MNRAS*, 294, 429
3. Dubner, G. M.; Holdaway, M.; Goss, W. M.; Mirabel, I. F. 1998, *ApJ*, 116, 1842
4. Farnes, J. S.; Gaensler, B. M.; Purcell, C.; Sun, X. H.; Haverkorn, M.; Lenc, E.; O'Sullivan, S. P.; Akahori, T. 2017, *MNRAS*, 467, 4777
5. Fromm, C. M., Perucho, M., Mimica, P., Ros, E. 2012, *A&A*, 588, A101

6. Goodall, P. T., Alouani-Bibi, F., Blundell, K. M. 2011, *MNRAS*, 414, 2838
7. Keppens, R., Meliani, Z., van Marle, A. J., Delmont, P., Vlasis, A., van der Holst, B. 2012, *JCoPh*, 231, 718
8. Mathews, W. G. 1971, *ApJ*, 165, 147
9. Monceau-Baroux, R., Porth, O., Meliani, Z., Keppens, R. 2014, *A&A*, 561, A30
10. Monceau-Baroux, R., Porth, O., Meliani, Z., Keppens, R. 2015, *A&A*, 574, A143
11. Porth, O., Xia, C., Hendrix, T., Moschou, S. P., Keppens, R. 2014, *ApJS*, 214, 4

Knots in Relativistic Transverse Stratified Jets

Z. Meliani and O. Hervet

1 Introduction

There is growing evidence of transverse stratification of relativistic astrophysical jets with clear indication of a fast inner jet (spine) embedded in a slower outer flow (layer). In many active galactic nuclei (AGN) a limb-brightened jet morphology is observed on pc scales [6], which is interpreted as an outcome of the differential Doppler boosting between the jet spine and layer [5]. This scenario is supported by multiple radio-load AGN observations, such as the well-known radio galaxy M87 which presents hints of a two-component jet from its polarized emission [1].

Numerous standing and moving radio knots in AGN jets have been observed over the last decades in very long baseline interferometry (VLBI) thanks to dedicated long-term monitoring programs such as MOJAVE[1] or TANAMI.[2] From these observations, stationary knots are often interpreted as re-collimation shocks resulting from the propagation of overpressured super-Alfvenic jets through the external medium. This pressure difference between the jet and the external medium is caused by the large distances covered by relativistic AGN jets in the galactic

[1] http://www.physics.purdue.edu/MOJAVE/

[2] http://pulsar.sternwarte.uni-erlangen.de/tanami/

Z. Meliani (✉)
LUTH, Observatoire de Paris, CNRS UMR 8102, Université Paris-Diderot, Meudon, France
e-mail: zakaria.meliani@obspm.fr

O. Hervet
Department of Physics, Santa Cruz Institute for Particle Physics, University of California at Santa Cruz, Santa Cruz, CA, USA
e-mail: ohervet@ucsc.edu

C. Sauty (ed.), *JET Simulations, Experiments, and Theory*,
Astrophysics and Space Science Proceedings 55,
https://doi.org/10.1007/978-3-030-14128-8_12

medium. Indeed, with distance the external medium pressure decreases faster than the jet pressure, which gives rise to rarefaction waves and shock waves within the jet. This phenomena was studied using the characteristic methods [4].

In order to study the effects of the jet transverse stratification, we elaborate a two-component jet model according to the jet formation scenarios [2, 3, 10] and numerical simulation [9]. We adopt a two-component jet model with various kinetic energy flux distributions between the inner-outer jets [7]. We aim to investigate how this energy distribution influences the overall jet stability, the re-collimation shock development, and the local jet acceleration. Also, we assume various configurations. The first case is an overpressured uniform jet propagating in the external medium. In all the other investigated cases, we set a transverse structured jet with an overpressured inner jet. This assumption results from the intrinsic properties of the launching region, as described above.

2 Model

To study re-collimation shocks in transverse stratified jets, we adopt a two-component jet model with two uniform components. The model uses the basic characteristics of relativistic AGN jets, such as the total kinetic luminosity flux within an interval $L_{\rm k} = \left[10^{43}\,,\, 10^{46}\right]$ ergs/s [11], and the outer radius of the two-component jet $R_{\rm out} = R_{\rm jet} \sim 0.1$ pc at a parsec scale distance from the black hole. For the less constrained inner jet radius, we adopt the initial value $R_{\rm in} = R_{\rm jet}/3$. As initial condition for simulations, we establish a cylindrical flow column along the jet axis with a radius $R_{\rm jet}$. Two types of jets are investigated in this paper, uniform jets (the reference case) and two-component jets. For structured jets, we have a discontinuity in the density, pressure, and velocity at the interface of the two components $R_{\rm in}$. The jet properties are related to the external medium density and pressure by

$$\rho = \begin{cases} \rho_0\ \ \eta_{\rho,\rm in} & R \leq R_{\rm in}\,, \\ \rho_0\ \ \eta_{\rho,\rm out} & R_{\rm in} < R < R_{\rm jet}\,, \end{cases} \quad \text{and } p = \begin{cases} p_0\ \ \eta_{p,\rm in} & R \leq R_{\rm in}\,, \\ p_0\ \ \eta_{p,\rm out} & R_{\rm in} < R < R_{\rm jet}\,. \end{cases} \tag{1}$$

where ρ_0 and p_0 are respectively the density and the pressure of the external medium, $\eta_{\rho,\rm in}$ and $\eta_{\rho,\rm out}$ are the inner and outer jet density ratio relative to the external medium density, and $\eta_{p,\rm in}$ and $\eta_{p,\rm out}$ are the inner and outer jet pressure ratio relative to the external medium pressure.

Five cases (A, B, C, D, E, and F) are investigated. The first case (A) presents a uniform jet with a Lorentz factor $\gamma = 10$. All the others cases are two-component jet simulations; their order follows an increasing ratio of the outer/inner jets component kinetic powers. Hence, case (B) has the most powerful inner jet, carrying 95% of the total kinetic power. Cases (C) and (D) are set with inner and outer jet carrying

Table 1 Most relevant parameters for the two models investigated, the density ratio η_ρ, the Mach number $\mathcal{M}_c$, the contribution of each jet components to the total kinetic energy flux of the jet $L_{k,\,in}/L_{k,\,total}$, $L_{k,\,out}/L_{k,\,total}$. In addition to these values, the external medium has a normalized number density $\rho_0 = 1$ and a normalized pressure $p_0 \simeq 5 \times 10^{-2}$ or $p_0 \simeq 1 \times 10^{-3}$ (these two types of pressure are chosen according to the energy ratio between inner and outer jet components). The inner jet always presents an initial Lorentz factor of $\gamma = 10$, higher than that of the outer jet initialized at $\gamma = 3$. The outer jet is assumed to be in pressure equilibrium with the external medium $\eta_{p,\,out} = 1$, contrary to the inner jet which presents a larger pressure $\eta_{p,\,in} = 1.5$

	Inner jet		Outer jet		Structured jet		
Case	$\eta_{\rho\,,in}$	$\mathcal{M}_{c,\,in}$	$\eta_{\rho\,,out}$	$\mathcal{M}_{c,\,out}$	$L_{k,\,in}/L_{k,\,total}$	$L_{k,\,out}/L_{k,\,total}$	Two-component jet
A	4.5×10^{-4}	1.22			1	0.0	No
B	5×10^{-1}	4.34	5×10^{-6}	1.16	0.95	0.05	Yes
C	5×10^{-3}	1.22	5×10^{-1}	16.34	0.70	0.30	Yes
D	5×10^{-6}	1.22	1×10^{-1}	6	0.25	0.75	Yes
E	5×10^{-3}	1.22	5×10	19.0	5×10^{-3}	0.995	Yes
F	1×10^{-3}	0	5×10^{-2}	6.0	0	1	Yes

relatively the same order of kinetic power. In case (E), the jet is set with very powerful outer jet carrying 99,93% of the total kinetic power; this is also the only case where the outer jet is denser than the external medium. In the last case (F), the inner jet is empty and all energy is carried by the outer jet. All these cases are listed in Table 1. We should notice, that the increase in the outer jet component's kinetic energy flux, induces an increase of its density and thus a decrease in the sound speed, as result, the Mach number of the outer jet component in the cases C, D, E increases.

3 Results and Discussion

Our transverse structured jet model shows that energy distribution between the inner and the outer jet could be the key to jet classification (Fig. 1). Indeed, the energy distribution has a significant influence on the formation and the state of internal shocks and on the local jet acceleration. The model we propose here shows how the jet structure affects the development of the stationary and the non-stationary shocks observed in AGN jets. The jet structures modify the configuration of the internal shocks and the accelerations in the rarefaction regions. This structure can result from the jet launching mechanism.

We can classify the relativistic structured jets according to the transverse energy distribution between the two components as follows. This distribution affects the transverse variation Mach number since the density between the two components is related to the energy flux within each component.

The jets with low-energy outer jet (case B) show weak shocks. Moreover, the rarefaction waves are inefficient at accelerating the jet. The outer jet plays the role of a shear layer isolating the inner jet from the external medium. Moreover, the

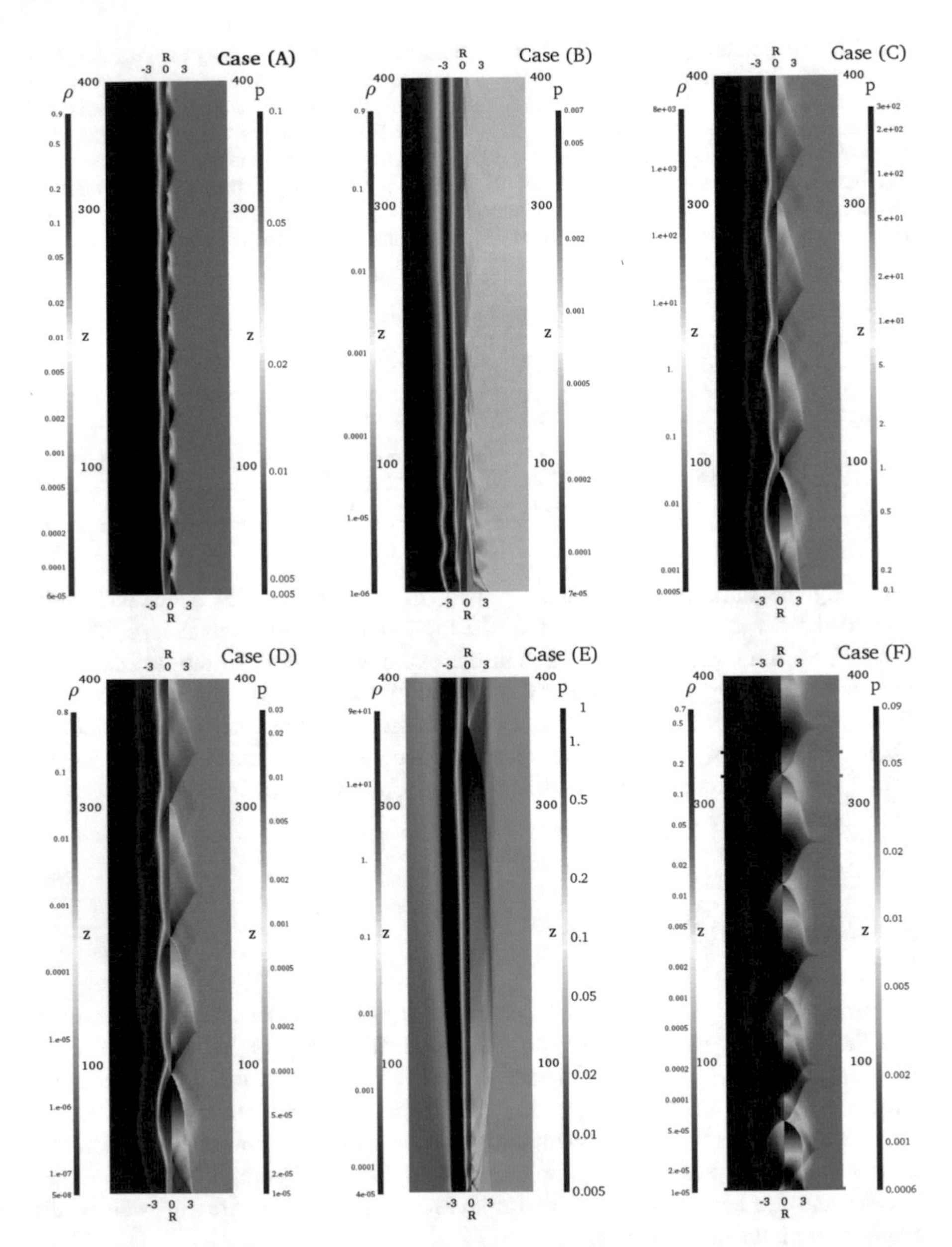

Fig. 1 Two-dimensional view of all the simulated cases of the jets along the poloidal direction. In each figure, the density color bar is drawn on the left side and the pressure color bar on the right side. The jet figures are stretched in the radial direction and squeezed in the longitudinal direction.The distance R and Z and in unit of the inner jet radius

low inertia of the outer component allows it to absorb waves. This makes a more efficient energy transfer from the inner to the outer jet. The Lorentz factor of the outer component increases with distance. Overall, the inner jet's Lorentz factor remains near the initial values.

The jets with near-equal energy distribution between the two components (cases C and D) show two shock wave structures with different wavelengths.

In case (D), the Lorentz factor reaches locally $\gamma \sim 30$ and even $\gamma \sim 50$. This acceleration is the result of the energy transfer from the outer to the inner jet by the inward propagating rarefaction waves that rise at the edge of the outer jet. The large difference in the Mach number between the hot inner jet and cold outer jet increases the efficiency of the energy transfer from outer to inner jet.

The jet with large energy carried by the outer jet, such as in case (E) (Fig. 1), could be representative of a jet with steady knots near the core and moving features at large distances like those observed in some sources. In the region with the steady shocks, the jet radius remains relatively constant, but downstream this radius increases with distance. The jet expansion at large distance is the result of the large inertia of outer jet that propagates in a rarefied external medium.

The last case (F) shows that jets with empty spines could evolve to conical shapes under the influence of the internal shocks.

These simulations show that the transverse structure in relativistic jets could be responsible for the diversity in knots observed in radio sources.

Acknowledgements Part of this work was supported by the PNHE and Observatoire de Paris. This work acknowledges financial support from the UnivEarthS Labex program at Sorbonne Paris Cité (ANR-10-LABX-0023 and ANR-11-IDEX-0005-02). O.H. thanks the U.S. National Science Foundation for support under grant PHY-1307311 and the Observatoire de Paris for financial support with ATER position. All the computations made use of the High Performance Computing OCCIGEN and JADE at CINES within the DARI project c2015046842. This research has made use of data from the MOJAVE database that is maintained by the MOJAVE team [8].

References

1. Attridge, J. M., Roberts, D. H. & Wardle, J. F. C. (1999), ApJ, 518, L87
2. Blandford, R. D., & Payne, D. G. (1982), mnras, 199, 883
3. Blandford, R. D., & Znajek, R. L. (1977), mnras, 179, 433
4. Daly, R. A. & Marscher, A. P. (1988), apj, 334, 539
5. Giovannini, G. (2003), NewAR, 47, 551
6. Giroletti M., Giovannini G., Feretti L., Cotton W.D., Edwards P.G., Lara L., Marscher A.P., Mattox J.R., Piner B.G. & Venturi T. (2004), ApJ, 600, 127
7. Hervet O., Meliani Z., Zech A., Boisson C., Cayatte V., Sauty C., and Sol H. (2017), A&A, 606, 103
8. Lister, M. L., Cohen, M. H., Homan, D. C., et al. (2009), aj, 138, 1874
9. McKinney, J. C. & Blandford, R. D. (2009), mnras, 394, L126
10. Meliani, Z., Sauty, C., Vlahakis, N., Tsinganos, K. & Trussoni, E. (2006), A&A, 447, 797
11. Rawlings, S., & Saunders, R. (1991), Nature, 349, 138

Part III
Observations and Experiments

ALMA Polarimetric Studies of Rotating Jet/Disk Systems

F. Bacciotti, J. M. Girart, M. Padovani, L. Podio, R. Paladino, L. Testi, E. Bianchi, D. Galli, C. Codella, D. Coffey, C. Favre, and D. Fedele

1 Introduction

The process of formation of stars and planets is one of the most intriguing topics in current astrophysics. In recent years, high angular resolution studies like the ones conducted with the Hubble Space Telescope (HST) and the Atacama Large Millimeter/submillimeter Array (ALMA) have allowed us to advance significantly in the knowledge of protoplanetary disks and associated outflows. In particular, the

F. Bacciotti (✉) · M. Padovani · L. Podio · D. Galli · C. Codella · F. Favre · D. Fedele
Istituto Nazionale di Astrofisica – Osservatorio Astrofisico di Arcetri, Firenze, Italy
e-mail: fran@arcetri.astro.it; padovani@arcetri.inaf.it; lpodio@arcetri.inaf.it; galli@arcetri.inaf.it; codella@arcetri.inaf.it; cfavre@arcetri.inaf.it; fedele@arcetri.inaf.it

J. M. Girart
Institut de Ciències de l'Espai (ICE, CSIC), Can Magrans, Catalunyia, Spain
e-mail: girart@ice.cat

R. Paladino
Istituto Nazionale di Astrofisica – Istituto di Radioastronomia Via P. Gobetti, Firenze, Italy
e-mail: paladino@ira.inaf.it

L. Testi
European Southern Observatory, Garching bei München, Germany
e-mail: ltesti@eso.org

E. Bianchi
Institut de Planétologie et d'Astrophysique de Grenoble (IPAG), Université Grenoble Alpes, Grenoble, France
e-mail: eleonora.bianchi@univ-grenoble-alpes.fr

D. Coffey
School of Physics, University College Dublin, Belfield, Ireland
e-mail: deirdre.coffey@ucd.ie

C. Sauty (ed.), *JET Simulations, Experiments, and Theory*, Astrophysics and Space Science Proceedings 55,
https://doi.org/10.1007/978-3-030-14128-8_13

sensitivity of ALMA allowed us to study the physical properties of young systems with an unprecedent combination of spectral and spatial resolution. One of the principal aims of such studies is to correctly set the initial conditions for planet formation.

In this context, the determination of the disk magnetic configuration is of particular interest, as magnetic fields may be responsible for the extraction of the excess of angular momentum from the system. This can occur via magneto-rotational instabilities generating an effective viscosity for horizonthal transport [6]. Such instabilities, however, have proven to be ineffective for disk realistic conditions [5]. Alternatively, protostellar jets generated by magneto-centrifugal acceleration can transport angular momentum vertically along the ordered strong magnetic field attached to the star and the disk [7, 9, 21]. Magneto-centrifugal winds can originate from the star ('stellar winds'), the disk co-rotation radius ('X-winds') or from a wider range of disk radii (extended 'disk winds'). In any case proto-planetary disks are expected to be strongly magnetized, which would have fundamental implications for planetary formation and migration models [27].

The increase in sensitivity in mm-wave polarimetry has opened a new possibility to investigate the disk properties, and in particular the magnetic configuration. Polarimetry, in fact, has long been believed to provide the orientation of magnetic field lines, as non-spherical dust grains tend to align with their short axis perpendicular to the direction of the magnetic field ('grain alignment'), giving rise to linear polarization of the emission [2].

Polarization, however, can also arise from self-scattering of the thermal emission of dust grains of size of the order of the radiation wavelength. In this case the models show that the polarization direction is parallel to the minor axis for inclined disks [17, 29].

A third effect producing polarized emission is the alignment of non-spherical grains with an anisotropic radiation field. For a centrally illuminated disk, the linear polarization would present for this mechanism a circular pattern centred on the source [25].

The first studies on protostellar envelopes allowed the identification of large-scale hourglass-shaped and twisted patterns consistent with the winding-up of magnetic field lines due to the rotation of the envelope [10, 22]. Subsequent studies at moderate resolution reported detections of polarized emission on protostellar discs (e.g. [16, 23]) but in none of these cases the polarization structure showed a clear relationship with the expected magnetic configuration. More recently, the inner disk scales have been reached thanks to the advent of ALMA. Maps of the polarization of the dust emission have been derived for various targets, with resolution down to $0.''1$–$0.''2$ (12–15 AU in nearby forming systems). A polarisation level of 0.5–2% can easily be detected over the relevant regions of the disk with the expected ALMA sensitivity. These studies show that all the mechanisms mentioned above can produce polarization, but dust self-scattering appears to be dominant [1, 11, 13, 17, 18, 24].

In this framework, we have started a project to map the polarization properties of young evolved Class II systems with associated jets, as these systems offer fundamental observational constraints in both the cases in which scattering or magnetic properties dominate the polarized emission. In particular, we are interested in the sources for which the rotation kinematics of both the jet and the disk has been studied. The determination of the magnetic configuration for these targets can constitute the ultimate proof of the validity of the magneto-centrifugal mechanism for the launch of jets, which can realize the extraction of the excess angular momentum.

Following this line of research we selected and recently observed with ALMA the Class II sources DG Tau and CW Tau. These targets have been the subject of numerous studies in the past, both for their disks and their collimated jets [3, 8, 12, 14, 19, 26]. The two sources are nearby (d$\sim$140 pc), free of their parental envelope, are oriented favourably for polarization measurements and their dust emission is sufficiently strong. Importantly, the known kinematics of the jets and disks allows one to identify immediately the near-side of the disks and to give constraints on the expected magnetic configuration.

Despite our expectations, the observed polarization (shown in Sect. 2) does not seem to convey a simple interpretation in terms of an ordered magnetic structure. Instead, the results are fully consistent with self-scattering of the dust emission (see discussion in Sect. 3). This allows us to derive from the polarization measurements new information on the size and distribution of dust grains in the disk [4].

2 Polarized Emission from DG Tau and CW Tau

In the following we illustrate the main features emerged from the polarimetric observations of DG Tau and CW Tau. More details can be found in [4].

The two targets were observed in full polarization mode in July 2017 within the ALMA Cycle 3 in Band 7 (870 µm). The configuration of the interferometer for these observations included 40 antennas, giving an angular resolution of about $0.''2$. The datasets were reduced and analysed with the Common Astronomical Software Application (CASA) software.

From the Stokes I, U, Q maps we obtain the linear polarization intensity, $P = \sqrt{Q^2 + U^2}$, the linear polarization fraction, $p = P/I$, and the polarization angle, $\chi = 0.5\,\arctan(U/Q)$, i.e. the direction of polarization of the electric field.

2.1 DG Tau

The case of DG Tau is illustrated in Fig. 1. The integrated flux is 880.2 ± 9.4 mJy, with peak intensity of 182.4 ± 1.4 mJy beam^{-1}, The FWHM along the major and minor axis are $0.''45$ and $0.''36$, respectively. These values imply $i_{\rm disk} \sim 37°$, while the measured disk PA is $135.4° \pm 2.5°$, almost perpendicular to PA$_{\rm jet} = 46°$.

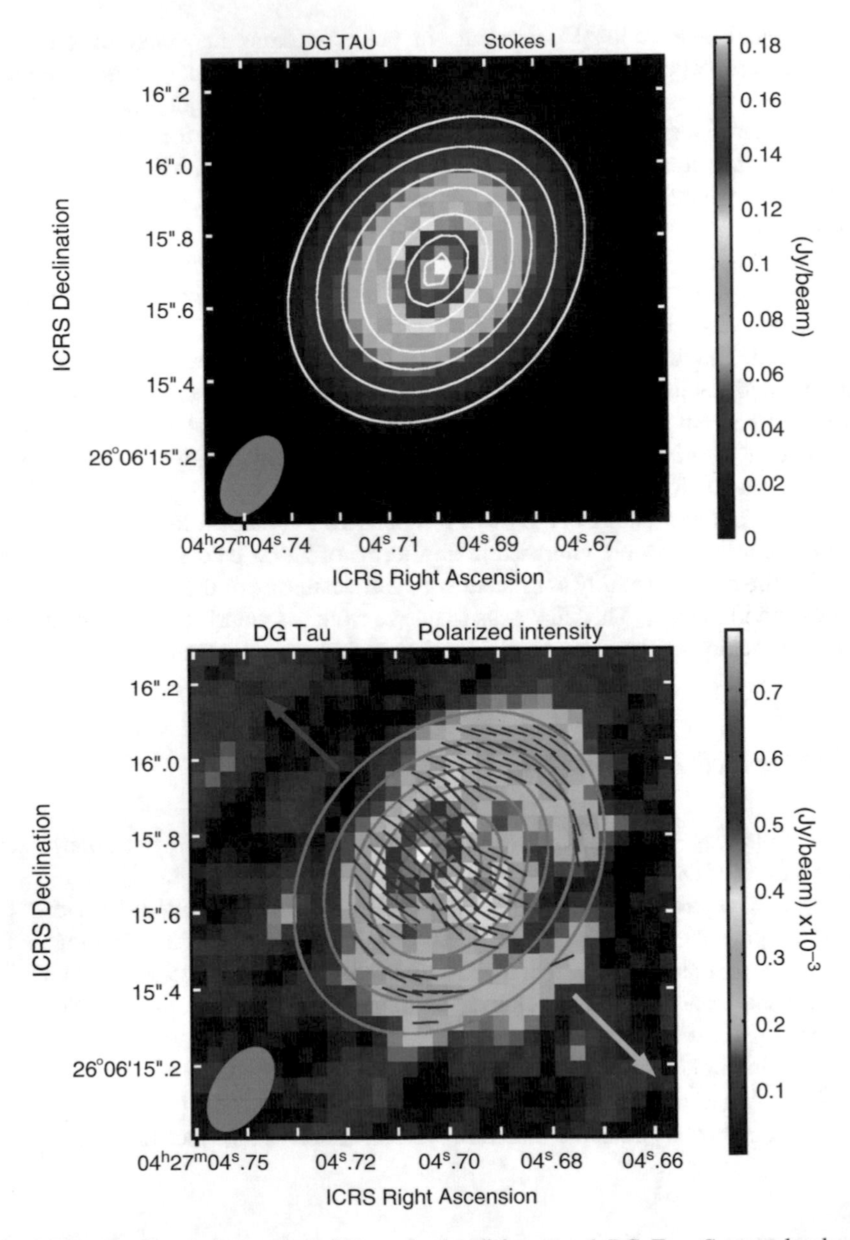

Fig. 1 *Top* Total emission map at 870 μm in the disks around DG Tau. Contour levels are [0.1, 0.2, 0.3, 0.4, 0.6, 0.8, 0.95] × peak value, which is 182.4 mJy beam^{-1} for DG Tau. *Bottom* Linearly polarized intensity P. Contours are as in top panel and the Polarization angle is indicated with vector bars. The arrows indicate the jet orientation, with the disk near-side on the same side of the red arrow. As discussed in Sect. 3, the evident asymmetry of the polarized intensity indicates a flared geometry for the disk

The polarization properties of DG Tau are illustrated in the bottom panel of Fig. 1. The near-side of the disk is brighter in polarized intensity, with the peak emission on the minor axis, displaced by $\sim$0.″07 from the photocenter of the total intensity. This feature, observed here for the first time in a disk around a low mass star, indicates that the disk has a flared geometry [29]. In the outer disk region, between 0.″3 and 0.″5 from the source, the polarized emission is distributed in a belt-like structure of lower intensity. The polarization vectors are nearly aligned with the disk minor axis in the central region, while they change orientation and become more azimuthal beyond 0.″3. The linear polarization fraction (not shown) reflects the distribution of the polarized intensity. Averaging over the whole disk area one finds $p = 0.41 \pm 0.17\%$.

2.2 CW Tau

Figure 2, top panel, illustrates the total intensity of the 870 μm continuum emission in CW Tau. For CW Tau, the integrated flux is 145.1 ± 1.4 mJy, with peak intensity of 44.9 ± 0.3 mJy beam^{-1}. The FWHM along the major and minor axis is 0.″35 and 0.″18 respectively. From these values, we estimate a disk inclination $i_{\rm disk}$ with respect to the line of sight of $\sim$59°. The disk PA is $60.7° \pm 1.9°$, almost perpendicular to the jet $\mathrm{PA_{jet}} = -29°$.

The bottom panel of Fig. 2 provides the map of the linearly polarized intensity P. This is centrally peaked and does not show any significant asymmetry. The polarization vectors are very well aligned along the minor axis of the disk. The polarization fraction, p, turns out to be almost constant in the disk central region of the disk, and it is on average $1.15 \pm 0.26\%$.

3 Dust Properties Derived from Polarization Measurements

The observed polarization properties can be explained in both sources in terms of self-scattering of the thermal dust emission [15, 28, 29]. This is in agreement with the findings in other protoplanetary disks [11, 13, 17]. If self-scattering dominates the polarization, no information can be gathered on the orientation of the magnetic field. This unfortunate occurrence, however, is partly compensated by the fact that the comparison of the observed maps and the predictions of the models for self-scattering give constraints on the size of th dust population of particles and on the geometry the disk.

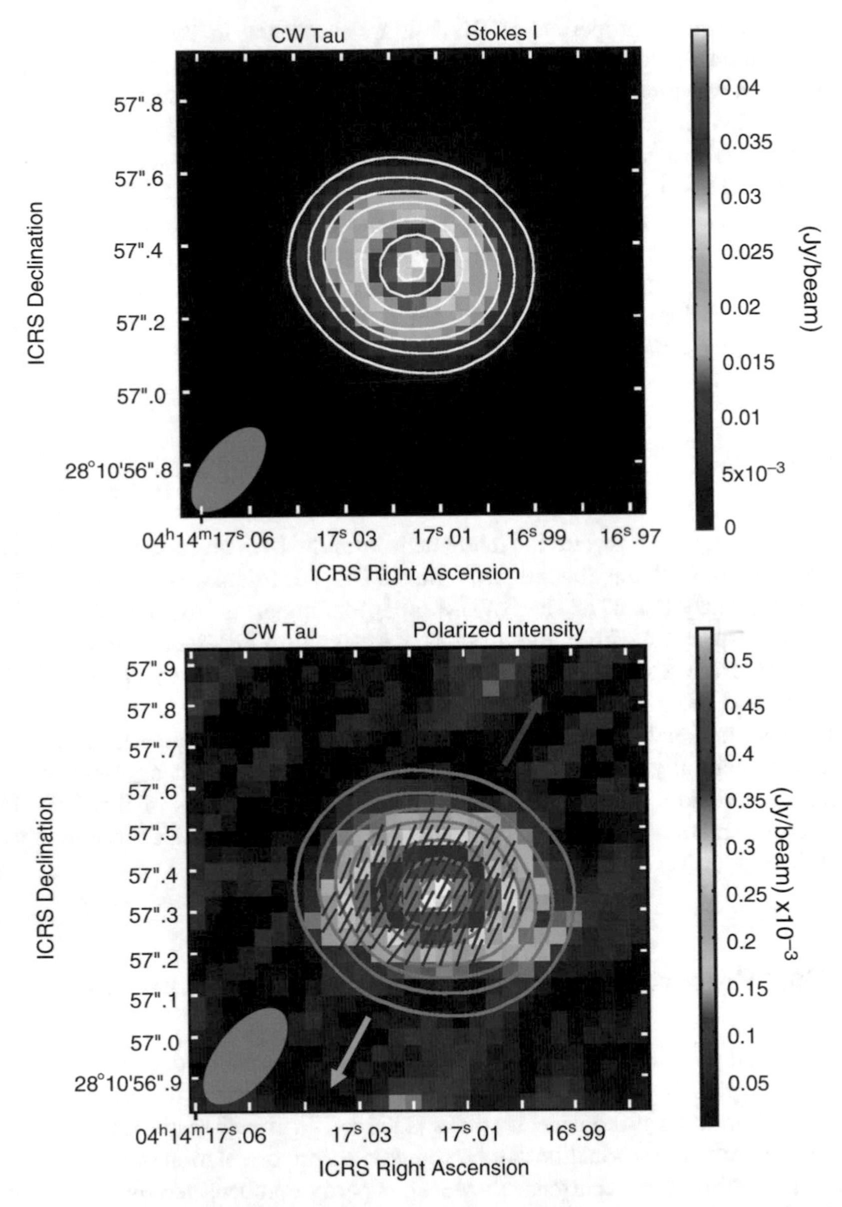

Fig. 2 *Top* Total emission map at 870 μm in the disks around CW Tau. Contour levels are [0.1, 0.2, 0.3, 0.4, 0.6, 0.8, 0.95] × peak value, which is 44.9 mJy beam^{-1} for CW Tau. *Bottom* Linearly polarized intensity P. Contours are as in top panel. polarization angle, χ, is indicated with fixed-length vector bars. The arrows follow the jet orientation. Disk near-side lies towards the receding jet lobe (red arrow). The central symmetry of the polarized intensity suggests a geometrically thin disk (see Sect. 3)

3.1 Grain Size

The analysis in [15] indicates that the maximum grain size contributing to polarization from self-scattering at a given wavelength, λ, is comparable to $\lambda/2\pi$. This implies a distribution peaked around 140 µm in our case. Further constraints can come from the diagnostic diagrams that correlate grain size, wavelength and polarization fraction, as investigated, e.g. in [17]. Using our values of the average polarization fraction and the diagrams in this work we estimate that the maximum grain size giving rise to the observed polarization is in the range 50–70 µm for DG Tau and about 100 µm for CW Tau (see Fig. 3).

3.2 Grain Settling Toward the Disk Midplane

In CW Tau, the distribution of the polarized intensity is symmetric, and polarization vectors are nearly parallel to the minor axis, with no curvature toward the outer disk. Since the disk around CW tau is reported to be optically thick by [19], the polarization models indicate that the observed features are consistent with self-scattering from a geometrically thin disk. Thus our observations indicate that the relatively larger grains in CW tau (see Section above) are more settled close to the disk midplane.

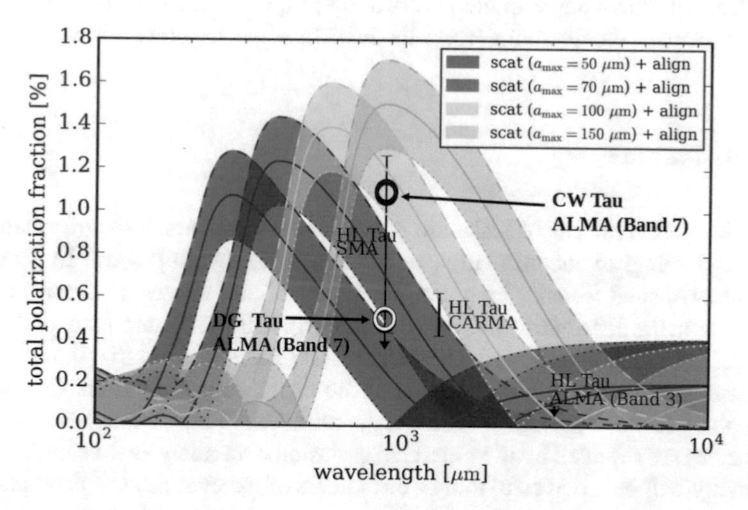

Fig. 3 Diagnostics of grain size using a diagram adapted from [17]. The vertical bars refer to past observations for the disk around HL Tau with various interferometers. Our observations, performed at 870 µm in ALMA Band 7, are indicated by the the black and white circles. Their position correspond to the average polarization fraction we measured for CW Tau and DG Tau, respectively. The average grain size turns out to be larger for CW Tau (about 100 µm) than for DG Tau (50–70 µm)

On the contrary, in DG Tau the polarization angle alignment is accompanied by an asymmetry in polarization intensity. This combination is consistent with the expectations of models of self-scattering in disks of intermediate or high optical depth, and with a finite angular thickness [29]. The disk is moderately optically thick according to [14], and the polarization maps are in agreement with the model expectations of [29]. Thus the observed asymmetry indicates that the scattering grains have not yet settled to the midplane.

3.3 Hints for Substructures in DG Tau?

We now consider the outer region of the DG Tau disk, i.e. beyond $0.''3$ from the source. The bottom panel of Fig. 1 shows structures in the polarized emission which do not correspond to any feature in the total intensity at the same resolution. In addition, a change in the orientation of the polarization pattern is observed. A possible explanation may come from a drop in the optical depth at $0.''3$ from the star, corresponding to about 45 au at the distance of the system. This may imply that there is a substructure in the disk density at this location, like a gap or a ring, not revealed in the total emission (see discussion in [4]). The nature of such a structure has still to be revealed, and will require higher angular resolution observations. We anticipate, however, that a recent study in the emission of molecular lines has revealed a ring in the emission of formaldheide (H2CO) whose inner border is at $0.''3$ from the star, coincident with the change in the polarization properties [20]. Thus it appears that polarization maps nicely complement the investigations in total emission.

4 Conclusions

The ALMA observations of disks are providing new and precious information for the understanding of the formation of planets around young stars. In particular, the window opened recently by polarimetric capabilities allows us to set important constraints on the distribution and early evolution of the dust component of disks. As other systems recently observed, DG Tau and CW Tau show polarization properties at 870 μm dominated by self-scattering of the dust thermal emission. Overall, DG Tau appears to be in a less evolved state than CW Tau. In addition, structural peculiarities are revealed by the polarized emission. Our analysis thus indicates that polarimetry will be a powerful tool in the studies of the evolution of protoplanetary disks.

Acknowledgements This paper uses ALMA data from project ADS/JAO.ALMA 2015.1.00840.S. ALMA is a partnership of ESO (representing its member states), NSF (USA) and NINS (Japan), together with NRC (Canada), MOST and ASIAA (Taiwan), and KASI (Republic of Korea), in cooperation with the Republic of Chile. The Joint ALMA Observatory is operated by ESO, AUI/NRAO and NAOJ. FB wishes to dedicate this work to the loving memory of her mother-in-law Giovanna (Janet) Pesenti. FB also wish thanks the editors for their understanding in a difficult time. Support is acknowledged from the project EU-FP7-JETSET (MRTN-CT-2004-005592).

References

1. Alves, F. O., Girart, J. M., Padovani, M., et al. Astron. Astrophys., **616**, A56 (2018)
2. Andersson, B.-G., Lazarian, A., Vaillancourt, J. E. Ann. Rev. Astron. Astrophys., **53**, 501 (2015)
3. Bacciotti, F., Ray, T. P., Mundt, R., Eislöffel, J., Solf, J. Astroph J. **576**, 222 (2002)
4. Bacciotti, F., Girart, J. M., Padovani, M., et al. Astroph. J. Lett. **865**, L12 (2018)
5. Bai, X.-N., & Stone, J. M. Astroph. J. **769**, 76 (2013)
6. Balbus, S. A., & Hawley, J. F. Astroph J. **376**, 214 (1991)
7. Blandford, R. D., Payne, D. G. Mon. Not. Roy. Astr. Soc. **199**, 883 (1982)
8. Coffey, D., Bacciotti, F., Ray, T. P., Eislöffel, J. and Woitas, J. Astroph. J. **663**, 350 (2007)
9. Frank, A., Ray, T. P., Cabrit, S., et al. in *Protostars and Planets VI*, ed. by H.Beuther, R.S. Klessen, C.P. Dullemond, and T.Henning, University of Arizona Press, Tucson, p.451–474 (2014)
10. Girart, J. M., Rao, R., Marrone, D. P. Science, **313**, 812 (2006)
11. Girart, J. M., Fernández-López, M., Li, Z.-Y., et al. Astroph J. Lett. **856**, L27 (2018)
12. Hartigan, P., Edwards, S., & Pierson, R. Astroph. J. **609**, 261 (2004)
13. Hull, C. L. H., Yang, H., Li, Z.-Y., et al. Astroph J. **860**, 82 (2018)
14. Isella, A., Carpenter, J. M., Sargent, A. I. Astroph. J. **714**, 1746 (2010)
15. Kataoka, A., Muto, T., Momose, M., et al. Astroph. J. **809**, 78 (2015)
16. Kataoka, A., Tsukagoshi, T., Momose, M., et al., Astroph. J. **83**1, L12 (2016)
17. Kataoka, A., Tsukagoshi, T., Pohl, A., et al. Astroph. J. Lett. **844**, L5 (2017)
18. Lee, C.-F., Li, Z.-Y., Ching, T.-C., Lai, S.-P., Yang, H. Astroph. J. **854**, 56 (2018)
19. Piétu, V., Guilloteau, S., Di Folco, E., Dutrey, A., Boehler, Y. Astron. Astrophys. **564**, A95 (2018)
20. Podio, L. et al., Astron. Astrophys. Lett., submitted
21. Pudritz, R. E., Ouyed, R., Fendt, C., & Brandenburg, A. in *Protostars and Planets V*, ed by B. Reipurth, D. Jewitt, and K. Keil, University of Arizona Press, Tucson, p.277–294 (2007)
22. Rao, R., Girart, J. M., Marrone, D. P., Lai, S.-P., Schnee, S. Astroph. J. **707**, 921 (2009)
23. Rao, R., Girart, J. M., Lai, S.-P., Marrone, D. P. Astroph. J. Lett. **780**, L6 (2014)
24. Stephens, I. W., Yang, H., Li, Z.-Y., et al. Astroph. J. **851**, 55 (2017)
25. Tazaki, R., Lazarian, A., & Nomura, H. Astroph. J. **839**, 56 (2017)
26. Testi, L., Bacciotti, F., Sargent, A. I., Ray, T. P., Eislöffel, J. Astron. Astrophys. **394**, L31 (2002)
27. Turner, N. J., Fromang, S., Gammie, C., et al., in *Protostars and Planets VI*, ed. by H.Beuther, R.S. Klessen, C.P. Dullemond, and T.Henning, University of Arizona Press, Tucson, p.411–432 (2014)
28. Yang, H., Li, Z.-Y., Looney, L., Stephens, I. Mon. Not. Roy. Astr. Soc. **456**, 2794 (2016)
29. Yang, H., Li, Z.-Y., Looney, L. W., Girart, J. M., Stephens, I. W. Mon. Not. Roy. Astr. Soc. **472**, 373 (2017)

Analysis of the Physical Properties of Jets/Outflows in T Tauri Stars

Fatima Lopez-Martinez and Jorge Filipe Gameiro

1 Introduction

It is known that the optical [O I] (λλ6300, 6363 Å), [N II] (λλ6548, 6583 Å) and [S II] (λλ6716, 6730 Å) forbidden lines are very sensitive tracers of outflows (see, e.g. [1, 4]). They are slightly affected by the extinction and their ratios do not depend on the geometry of the system as they are optically thin, allowing us to measure the physical properties of the formation region where they are formed. These forbidden lines in T Tauri stars (TTS) are characterized in many cases by two components: a high velocity component (HVC) and a low velocity component (LVC). The HVC is connected with extended collimated jets, whereas the origin of the outflows traced by the LVC is not yet fully understood. It could be emitted at the base of a magnetically driven disk wind [4] or could be a tracer of photoevaporative disk wind [2] or bound gas disk [7]. For a better understanding of the LVC origin in TTS a more suitable study of their properties is needed.

In [5] we developed an IDL based code to determine the electron temperature (T_e) and density (n_e) of the region where the semi-forbidden C II], Si II] and Fe II] UV lines are formed. The code found the best fitting to the spectra using the theoretical lines flux ratios, which yields reliable estimations of T_e and n_e of the emitting region. Our results pointed out that line emission is dominated by the magnetospheric accretion flow, close to the disk. We also found three sources with blueshifted line centroids, likely due to their jets and outflows, suggesting that the properties in the base of the outflow and at the base of the accretion flow are similar. To go into detail about these similarities, we adapted the UV analysis to the optical forbidden lines [O I], [N II] and [S II] in order to determine the properties of the

F. Lopez-Martinez (✉) · J. Filipe Gameiro
Instituto de Astrofísica e Ciências do Espaço, Universidade do Porto, Porto, Portugal
e-mail: Fatima.Lopez@astro.up.pt; jgameiro@astro.up.pt

C. Sauty (ed.), *JET Simulations, Experiments, and Theory*,
Astrophysics and Space Science Proceedings 55,
https://doi.org/10.1007/978-3-030-14128-8_14

emitting region. In addition, we analyzed the UV semiforbidden lines of C II], Fe II] and Si II] around 2326 Å at the base of their jets and we compared them with the optical results.

2 Method

In this work we adapted the code from [5] to the forbidden optical lines: [O I], [N II] and [S II]. In order to derive the electron temperature we used the known sensitivity of the flux ratio $[S II]_{6730}/[S II]_{6716}$ to it from $\sim 10^2$ cm^{-3} up to $\sim 10^5$ cm^{-3} [6], as well as the flux ratio $[O I]_{6300}/[S II]_{6730}$, which is sensitive to the n_e from $\sim 10^4$ to $\sim 10^7$ cm^{-3}. For the electron temperature we used different emission ratios. The combined analysis of all these ratios allow us to determine an unique temperature and density for the emission region. We assumed all the lines are optically thin and they are formed via collisional excitation in a single plasma characterized by a temperature T_e and a density n_e. All flux line ratios relative to the [S II] at 6730 Å were computed for a grid of temperatures from 10^4 to $10^{6.275}$ K in steps of 0.025 dex in $\log(T_e)$ and densities from 10^1 to $10^{9.25}$ cm^{-3} in steps of 0.25 dex in $\log(n_e)$. Then, for each pair (T_e, n_e) of the grid we found the best fit (i.e. with the minimum χ^2) of the observed [O I], [N II] and [S II] lines with one or two Gaussian functions (no more were needed) for each line. When two components were fitted to a given line we applied the same procedure for each component, i.e. we used two similar and independent grids of T_e and n_e to each component. Then the fit is made simultaneously for all components and lines.

3 Results and Data Analysis

We applied our method to 4 TTS with known jets: DG Tau, CW Tau, RW Aur and SZ 102. We used five observations carried out with the slit perpendicular to the jet direction on the base of SZ 102's receding jet, SZ 102's approaching jet, RW Aur's receding jet, DG Tau's approaching jet and CW Tau's approaching jet. The data were taken from the *Hubble Space Telescope* (HST)[1]/STIS data archive (ID 9435). We also analyzed a SZ 102 observation on the star taken from the European Southern Observatory (ESO) archive (ID: 71.C-0429(C)), and DG Tau and CW Tau star's observation (observed by our group on the WHT/UES). The results for T_e, n_e, FWHM and peak velocities yielded by the code are shown in Table 1.

[1]Based on observations made with the NASA/ESA Hubble Space Telescope, obtained from the data archive at the Space Telescope Science Institute. STScI is operated by the Association of Universities for Research in Astronomy, Inc. under NASA contract NAS 5-26555.

Table 1 Results corresponding to the best fitting

	LVC				HVC			
Star	$\log(T_e)$ (K)	$\log(n_e)$ (cm^{-3})	δ[1] ($km\ s^{-1}$)	FWHM ($km\ s^{-1}$)	$\log(T_e)$ (K)	$\log(n_e)$ (cm^{-3})	δ[1] ($km\ s^{-1}$)	FWHM ($km\ s^{-1}$)
SZ102/star[2]	4.550	2.25	−5(±4)[3]	275(±4)[3]	–	–	–	–
SZ102/star[2]	4.175	4.25	16(±4)[3]	113(±4)[3]	–	–	–	–
SZ102 Recd.	4.200	4.00	13(±16)	139(±36)	4.250	5.00	77(±28)	63(±38)
SZ102 Appr.	4.225	4.25	20(±25)	167(±39)	4.300	5.25	−113(±60)	198(±60)
DGTau/UES	5.475	5.75	−60(±10)	114(±24)	4.20	5.00	−178(±38)	190(±32)
DGTau Appr.	5.600	6.50	−61(±15)[3]	86(±24)	4.20	5.25	−183(±15)[3]	137(±28)
RWAur Recd.	–	–	–	–	4.125	3.25	91(±15)[3]	67(±15)[3]
CWTau/star	5.425	6.25	−4(±3)[3]	35(±3)[3]	5.250	6.75	−101(±6)	82(±6)
CWTau Appr.	5.600	6.75	−15(±15)	68(±28)	4.200	4.00	−111(±15)[3]	84(±15)[3]

The error bars correspond to the procedure uncertainties
[1]Line velocity shift in the stellar rest frame
[2]This star shows two LVC
[3]The error bars correspond to the resolution in velocity of the instrument since the procedure uncertainties were lower in these cases

Fig. 1 Results of temperatures and densities corresponding to the best fits for all the observations. Black, blue and red symbols represent star, approaching and receding jets observations, respectively. sz corresponds to SZ 102, dg to DG Tau, rw to RW Aur and cw corresponds to CW Tau

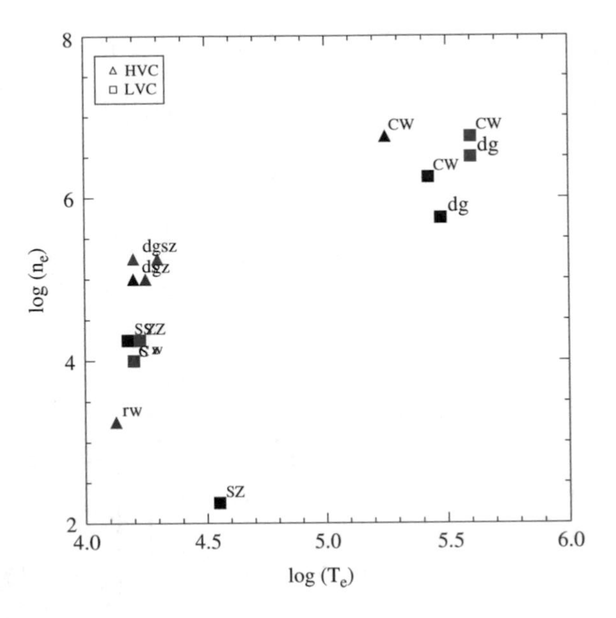

For the analysis of the fitted components to the line profiles we used $70\,\mathrm{km\,s^{-1}}$ as the threshold to separate LVC from HVC. The physical properties for jet observations presented in this work were derived from their spectra integrated over the jet width (the slit was located parallel to the disk at 0.″3 and 0.″2 from the star). Thus, the T_e and n_e should be understood as average values over the emitting region.

In Fig. 1 we plot all the derived T_e and n_e values. Two different ranges of T_e and n_e for the emitting regions are well identified, a group with $4.125 \leq \log T_e(\mathrm{K}) \leq 4.55$ and $2.25 \leq \log n_e(\mathrm{cm}^{-3}) \leq 5.25$, and another one with $5.25 \leq \log T_e(\mathrm{K}) \leq 5.6$ and $5.25 \leq \log n_e(\mathrm{cm}^{-3}) \leq 6.75$, with a clear general trend: the higher the temperature the denser the emitting region. For DG Tau and CW Tau, the LVC is emitted in a region with high temperature and high density, whereas for SZ 102 the LVC is formed in a less hot and less dense region.

We explored the connection between accretion process and physical properties of the HVC and LVC emitting regions. For the HVC there was not observed any correlation statistically significant, which is expected if an episodic accretion occurs in the jet driving region. However for the LVC we found that the hotter the emitting gas, the higher the accretion luminosity. This trend suggests that the accretion process is the main responsible of the heating of the region where the LVC is formed. Moreover, the accretion is also related with the density of the LVC gas. We found denser LVC emitting regions in those systems with higher accretion luminosity. These two last relations show that there is a connection between accretion and outflow processes, which is expected if a steady accretion drives the outflow process. For both the LVC and the HVC we found: the lower the disk inclination, the higher the peak velocity, having the highest peak velocity the star seen more pole-on of the sample (DG Tau). We also studied the relation between the FWHM and the inclination of the system, finding that the higher the inclination of the system, the higher the FWHM of the LVC.

These results point out that the LVC comes from a disk wind, where Keplerian and likely expanding wind streamlines are responsible of the broadening of the lines. Assuming that the broadening of the lines is mainly due to Keplerian rotation, we computed that the LVC comes from a disk wind at 0.05–1.69 AU from the source.

3.1 UV Analysis

We analyzed two UV observations carried out with the slit perpendicular to the jet direction at 0.″3 from the source at the base of SZ 102's receding jet and DG Tau's approaching jet. These observations were extracted from the HST/STIS (E230M) archive (ID: 9807). We studied the UV lines of C II], Fe II] and Si II] in the spectral range of ∼2324–2380 Å. In these observations on the jets the emission from the Fe II] and Si II] lines is very weak, however the C II] quintuplet was detected in both observations, indicating that the C II] lines are also a good tracer of outflows [3]. We derived the physical properties of the emitting region following [5] and the results are shown in Table 2. For comparison, the properties derived in [5] for DG Tau and SZ 102 from UV observations on the star are also shown.

The T_e and n_e values determined for SZ 102 from UV data are very similar for both star and jet observations, suggesting that indeed the star observation analyzed in [5] was dominated by the outflow, where we found this low density. For the DG Tau approaching jet (UV observation) we obtained a very low density, which is far from the densities derived from the optical data for the same jet and star ($\log(n_e) > 5.0\,\mathrm{cm}^{-3}$). These differences could be due to the time elapsed between the observations, however we can also observe the two regions with different properties in DG Tau approaching UV and optical HST observations, that were taken simultaneously. Therefore, the analysis of both UV and optical lines will provide us the physical properties of different emitting regions of the outflows/jets.

Acknowledgements This work was supported by Fundação para a Ciência e a Tecnologia (FCT) through national funds (UID/FIS/04434/2013) and by FEDER through COMPETE2020 (POCI-01-0145-FEDER-007672). F.L-M. acknowledges the support by the IA fellowship CIAAUP-02/2016-BPD in the context of the project (UID/FIS/04434/2013 & POCI-01-0145-FEDER-007672).

Table 2 Results corresponding to the best fitting found for DG Tau and SZ 102 from C II], Fe II] and Si II] line ratios in the UV range

Star	$\log(T_e)$ (K)	$\log(n_e)$ (cm^{-3})	δ^1 ($\mathrm{km\ s}^{-1}$)	FWHM ($\mathrm{km\ s}^{-1}$)
SZ102/star[2]	4.45	1.25	20	275
SZ102/Recd.	4.32	1.75	25	143
DGTau/star[2]	4.15	10.0	−87	122
DGTau/Appr.	4.25	2.00	−188	128

[1]Corrected by the radial velocity of the system
[2]Results from [5]

References

1. Edwards, S., Cabrit, S., Strom, S. E., et al. 1987, ApJ, 321, 473
2. Ercolano, B., & Owen, J. E. 2016, MNRAS, 460, 3472
3. Gómez de Castro, A. I., & Ferro-Fontán, C. 2005, MNRAS, 362, 569
4. Hartigan, P., Edwards, S., & Ghandour, L. 1995, ApJ, 452, 736
5. López-Martínez, F., & Gómez de Castro, A. I. 2014, MNRAS, 442, 2951
6. Osterbrock, D. E. 1989, skytel, 78, 491
7. Simon, M. N., Pascucci, I., Edwards, S., et al. 2016, ApJ, 831, 169

X-Shooter Survey of Jets and Winds in T Tauri Stars

Brunella Nisini, Simone Antoniucci, Juan Manuel Alcalá, Teresa Giannini, Carlo Felice Manara, Antonella Natta, Davide Fedele, and Katia Biazzo

1 Introduction

It is a common paradigm that collimated jets are an ubiquitous phenomenon during the active phase of accretion, as they are able to efficiently extract angular momentum accumulated in the disc, allowing disc matter to accrete onto the star (e.g. [1]). Powerful optical jets are indeed commonly observed in young protostars and very active Classical T Tauri (CTT) stars. However, the frequency of the jet phenomenon among the general population of young stellar objects is still far from

B. Nisini (✉) · S. Antoniucci · T. Giannini
INAF – Osservatorio Astronomico di Roma, Rome, Italy
e-mail: brunella.nisini@inaf.it; simone.antoniucci@inaf.it; teresa.giannini@inaf.it

J. M. Alcalá
INAF – Osservatorio Astronomico di Capodimonte, Napoli, Italy
e-mail: juan.alcala@inaf.it

C. F. Manara
ESO Headquarters, München, Germany
e-mail: cmanara@eso.org

A. Natta
DIAS/School of Cosmic Physics, Dublin Institute for Advanced Studies, Dublin, Ireland

D. Fedele
INAF – Osservatorio Astrofisico di Arcetri, Firenze, Italy
e-mail: fedele@arcetri.inaf.it

K. Biazzo
INAF-Osservatorio Astrofisico di Catania, Catania, Italy
e-mail: katia.biazzo@inaf.it

C. Sauty (ed.), *JET Simulations, Experiments, and Theory*, Astrophysics and Space Science Proceedings 55, https://doi.org/10.1007/978-3-030-14128-8_15

being settled. The question on how common jets are in active accreting stars has a major relevance in the context of disc evolution, as jets can significantly modify the structure of the disc region involved in the launching, impacting its evolution and the formation of planetary systems.

A direct way to infer the presence of jets and winds in CTT stars is through observations of atomic forbidden lines (e.g. [2, 3, 5]), and of the [OI]6300Å transition in particular, being the brightest among the optical lines. The [OI]6300Å line commonly exhibits two distinct components: a Low Velocity Component (LVC), peaking at velocities close to systemic one or only slightly blue-shifted, and a High Velocity Component (HVC) observed at velocities up to $\sim \pm\, 200\,\mathrm{km\,s^{-1}}$. The HVC is directly connected with the extended collimated jets, while the origin of the LVC is still unclear. It comes from a compact and denser region with respect to the HVC [2, 3] and it is usually associated to an atomic photo-evaporated or magneto-hydro-dynamic (MHD) slow-wind (e.g. [6, 7]). In this contribution, we summarise the results obtained with a statistical study on the [OI]6300Å emission in a sample of 131 T Tauri stars in the Lupus, Chamaeleon and σ Orionis star-forming regions [4] observed with the X-shooter instrument at the VLT. These observations are part of a project aimed at characterizing the population of young stellar objects with discs in nearby star-forming regions [8, 9]. As part of this project, the stellar and accretion properties of the sample have been measured in a uniform and self-consistent way, providing a unique database where the properties of the [OI]6300Å emission can be analysed and correlated with the other derived parameters in a homogeneous fashion.

2 The Sample and [OI]6300Å Decomposition

The adopted sample consists of 131 Young Stellar Objects (YSOs) observed with the X-shooter instrument. In particular, the sample includes sources in the Lupus (82 sources), Chamaeleon (41 sources) and σ Orionis (8 sources) star forming regions, whose stellar and accretion properties have been already analysed in previous papers ([8, 10] for Lup; [11] for σ Ori; [9] for Cha). We focus our analysis on the [OI]6300Å line which is the brightest among the class II forbidden lines in the spectral range of our data. As discussed in the introduction, the [OI]6300Å profile can be decomposed in a LVC, close to systemic velocity, and a blue- or red-shifted HVC. To separately derive line center and width of the different components we fitted the [OI] line profile with one or more gaussians. We define the velocity that separates the LVC from the HVC as $40\,\mathrm{km\,s^{-1}}$, which roughly corresponds to the resolution attained for most of our objects. Line fluxes have been then converted into intrinsic luminosities adopting the A_V derived from the stellar parameters fitting procedure (see e.g. [10]).

3 Statistics and Correlation with Stellar and Disc Properties

The [OI]6300Å line is detected in 101 out of the 131 sources of our sample (i.e. 77% rate of detection). In all detections the LVC is identified, while the HVC is observed in 39/131 objects (i.e. 30% rate of detection). The $L_{[OI]\mathrm{HVC}}$ spans a wide range of values, i.e. between 0.01 and 3 $L_{\odot}$ for the detections. In the majority of the sources the $L_{[OI]\mathrm{HVC}}$ is typically fainter than the $L_{[OI]\mathrm{LVC}}$ by a factor of two/three, which partially explains the higher number of LVC detections with respect to the HVC. However sensitivity alone is not the only reason for a difference in detection rate between the LVC and HVC, as the signal-to-noise ratio in the LVC line is often larger than 10. We also estimate that orientation effects could misclassify emission at high velocity with a LVC in about 20% of sources (see discussion in [4]). The HVC detection rate does not depend on stellar luminosity, while a slightly larger detection rate is observed in sources with higher accretion luminosity (39% in sources with log $(L_{\mathrm{acc}}/L_{\odot}) > -3$). Our sample comprises also 11 transitional discs (TD) sources. The detection rates in these objects is roughly in line with the overall statistics of our sample (i.e. 80% LVC, 20% HVC), which confirms that TD have characteristics similar to other CTTs with a full disc, not only for what accretion properties concern [12], but also regarding the properties of the inner atomic winds.

The [OI]6300Å line luminosity has been correlated with the stellar and accretion parameters for both the LVC and HVC. The line luminosity correlates well with all the considered parameters in both LVC and HVC, with a Pearson correlation coefficient r always between 0.7 and 0.8. Correlations with L_* present the larger scatter and in general we find that both the LVC and HVC show a better correlation with L_{acc} than with all the other parameters. The best fit linear regressions for all

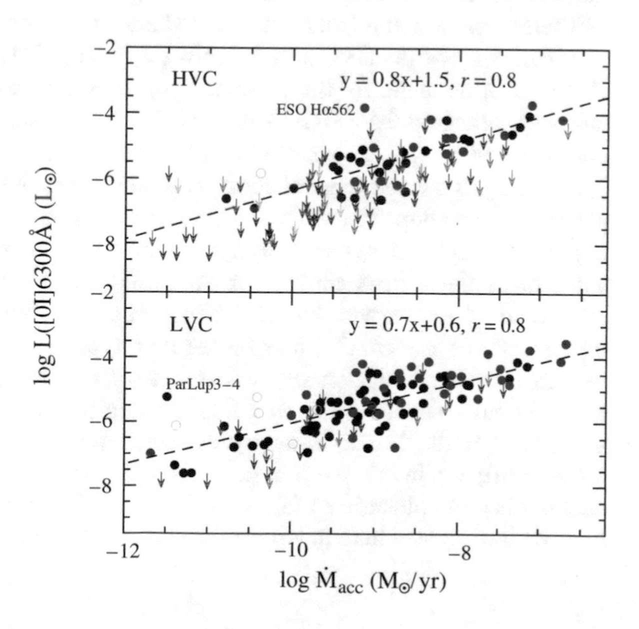

Fig. 1 The [OI]6300Å line luminosity of the LVC (bottom) and HVC (top) is plotted as a function of the mass accretion rate. Arrows refer to 3σ upper limits. The dashed line indicates the linear regression whose parameters are given in the upper right corner of the figure, together with the correlation parameter. Black, red and green symbols indicate sources in Lupus, Chamaelion and σ Orionis clouds, respectively. (From [4])

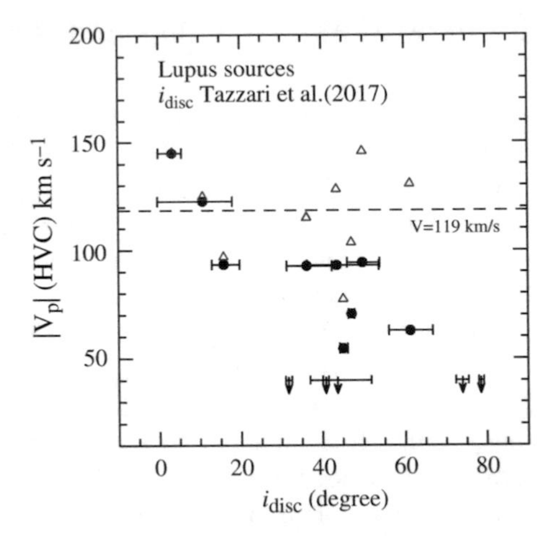

Fig. 2 The absolute value of the peak velocity in the HVC (V_{peak}(HVC)) is plotted as a function of the disc inclination angle for a sub-sample of sources in Lupus with known disk inclination (i_{disc}). An upper limit of V_{peak}(HVC) of 40 km s^{-1} has been applied for sources with measured inclination angles that don't have a HVC component. Red triangles refer to the total jet velocity obtained by de-projecting the peak radial velocity by the corresponding inclination angle for the sources with a detected HVC. The dashed line indicates the average jet total velocity of the sample. (Figure adapted from [4])

the correlations can be found in [4], while in Fig. 1 we show the correlation between the [OI]6300Å luminosity and the mass accretion rate $\dot{M}_{acc}$. The log($L_{[OI]\rm HVC}$) vs log $\dot{M}_{acc}$ correlation is particularly useful to indirectly estimate the mass accretion rate in moderately embedded sources driving jets, where the permitted lines or the LVC emission region might be subject to large extinctions or scattering (e.g. [4]).

Figure 2 shows the HVC peak velocity (V_{peak}(HVC)) plotted as a function of the disc inclination angle for the 13 sources of the Lupus cloud where this parameter has been measured from ALMA images [13]. The absolute value of V_{peak} is anti-correlated with i_{disc}, as expected in the working hypothesis that the HVC traces collimated jets ejected in a direction perpendicular to the disc. If we de-project the peak velocities adopting the corresponding inclination angle, we find total velocities ranging between ∼100 and 150 km s^{-1} (red triangles in Fig. 2), with an average value of 119 km s^{-1}. A similar correlation has been derived by [14] in a sample of T Tauri of Taurus with known inclination angle, showing a jet mean velocities of the order of 200 km s^{-1}, thus higher than typical values found by us. This can be explained by a larger mass, on average, of the sources in Taurus with respect to the Lupus sources, as the jet poloidal velocity is proportional to the Keplerian velocity v_K in the disc at the launching radius r_0 ($v_K = \sqrt{(GM_*/r_0)}$). Our derived jet velocities are in fact more in line with the velocities found from proper motion analysis in a sample of jets of the Cha II cloud [15], driven by sources more similar to those investigated here in terms of stellar parameters.

4 The $\dot{M}_{jet}/\dot{M}_{acc}$ Ratio

In the assumption that the [OI]6300Å HVC traces the high velocity jet, its luminosity can be used to infer the jet mass flux rate ($\dot{M}_{jet}$) adopting a procedure similar to that applied in [2]. In particular, to estimate the mass of the flow from the line luminosity we have considered the [OI]6300Å emissivities computed considering a 5 level OI model, and assuming $T_e = 10{,}000$ K and $n_e = 5\times10^4$ cm^{-3}, which typically represent the values found in the region at the base of T Tauri jets, where the [OI] emission originates. The adopted values are also similar to those used by [2] to estimate the $\dot{M}_{jet}$ of a sample of Taurus sources from their HVC [OI] emission. Hence, we can make a direct comparison between our derived mass flux rates and those estimated by [2].

In Fig. 3 the derived mass flux rates are plotted as a function of the source mass accretion rates. In this figure we also plot the data points relative to the sample of sources in Taurus whose $\dot{M}_{jet}$ has been measured by [2] with a similar procedure as the one we adopt here. For these sources, we have estimated the $\dot{M}_{acc}$ from the determination of the accretion luminosity given by [16], adopting the stellar masses and radii given in the same paper. The Taurus sources, which extend to higher mass accretion rates with respect to our sample, fall along the same trend as the rest of targets. The Figure shows that for most of the detections the $\dot{M}_{jet}/\dot{M}_{acc}$ ratio is between 0.01 and 0.5, confirming, on a large statistical bases, previous results found on individual objects. We note a significant number of upper limits pointing to a very low efficiency of the jet mass loss rate, even for objects with high accretion rates. In

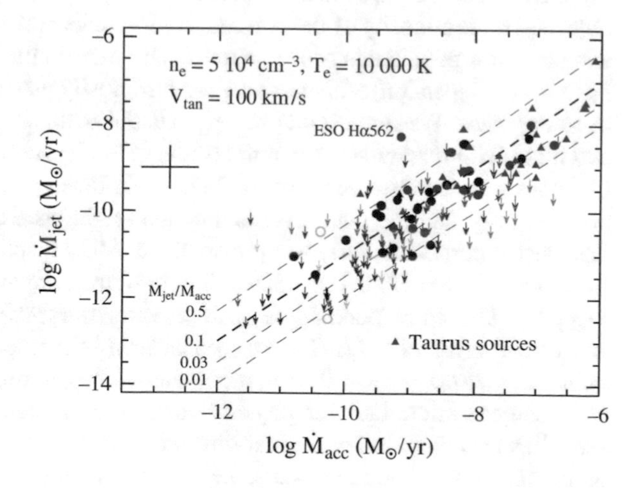

Fig. 3 Mass accretion ($\dot{M}_{acc}$) vs mass ejection ($\dot{M}_{jet}$) rates. The $\dot{M}_{jet}$ has been measured from the [OI]6300Å luminosity assuming a jet tangential velocity of 100 km s^{-1} and gas temperature and density of 10,000 K and 5×10^4 cm^{-3} respectively. Black, red and green symbols are as in Fig. 1, while blue triangles indicate the values for the sample of Taurus sources (see text). (From [4])

particular, the average of the detections is at $\dot{M}_{jet}/\dot{M}_{acc} = 0.07$ but considering both detections and upper limits, we find that 57 sources, i.e. 44% of the entire sample, have $\dot{M}_{jet}/\dot{M}_{acc} < 0.03$.

The large scatter in the $\dot{M}_{jet}/\dot{M}_{acc}$ ratio is partially due to the uncertainty in the determination of the $\dot{M}_{jet}$, and in particular on the assumed parameters. However, given the remarkable similarity in the excitation conditions of jets from sources with different masses and mass accretion rates, it is likely that the scatter, of almost two order of magnitude, reflects a real difference in the $\dot{M}_{jet}$ efficiency among sources more than excitation conditions very different from what assumed here.

In conclusion, the analysis of $\dot{M}_{jet}/\dot{M}_{acc}$ ratio shows that there is a large spread of values for this ratio that does not depend on the mass of the driving source. From Fig. 3 we note, however, a tentative trend for sources with accretion rates $>10^{-8}$ $M_{\odot}\,yr^{-1}$ to have on average a lower ratio. This trend needs to be confirmed/rejected on the bases of observations on a more complete sample of sources with high mass accretion rates.

5 Conclusions

We have summarised the results of a statistical work performed on the [OI]6300Å line in a sample of 131 young stars with discs in the Lupus, Chamaeleon and σ Orionis star forming regions, having mass accretion rates spanning from 10^{-12} to 10^{-7} $M_{\odot}\,yr^{-1}$, observed with the X-shooter instrument. The line has been deconvolved into a LVC, detected in 77% of the sources, and in a HVC, present in only 31% of the objects. The luminosity of the two line components correlates with different stellar and accretion parameters of the sources. The line luminosity correlates better (i.e. has a lower dispersion) with the accretion luminosity than with the stellar luminosity or stellar mass. We suggest that accretion is the main drivers for the line excitation and that MHD disc-winds are at the origin of both components. In the sub-sample of Lupus sources observed with ALMA a relationship is found between the HVC peak velocity and the outer disc inclination angle, as expected if the HVC traces jets ejected perpendicularly to the disc plane. Mass ejection rates ($\dot{M}_{jet}$) measured from the detected HVC [OI]6300Å line luminosity span from $\sim 10^{-13}$ to $\sim 10^{-7}$ $M_{\odot}\,yr^{-1}$. The corresponding $\dot{M}_{jet}/\dot{M}_{acc}$ ratio ranges from $\sim$0.01 to $\sim$0.5, with an average value of 0.07. However, considering the upper limits on the HVC, we infer a $\dot{M}_{jet}/\dot{M}_{acc}$ ratio <0.03 in more than 40% of sources. We argue that most of these sources might lack the physical conditions needed for an efficient magneto-centrifugal acceleration in the star-disc interaction region. Systematic observations of populations of younger stars, i.e. class 0/I, are needed to explore how the frequency and role of jets evolve during the pre-main sequence phase. This will be possible in the near future thanks to space facilities like JWST.

References

1. Frank, A., Ray, T. P., Cabrit, S., et al. 2014, Protostars and Planets VI, 451
2. Hartigan, P., Edwards, S., & Ghandour, L. 1995, ApJ, 452, 736
3. Natta, A., Testi, L., Alcalá, J. M., et al. 2014, A&A, 569, A5
4. Nisini, B., Antoniucci, S., Alcalá, J. M., et al. 2018, A&A, 609, A87
5. Simon, M. N., Pascucci, I., Edwards, S., et al. 2016, ApJ, 831, 169
6. Casse, F., & Ferreira, J. 2000, A&A, 353, 1115
7. Ercolano, B., & Owen, J. E. 2010, MNRAS, 406, 1553
8. Alcalá, J. M., Manara, C. F., Natta, A., et al. 2017, A&A, 600, A20
9. Manara, C. F., Testi, L., Herczeg, G. J., et al. 2017, A&A, 604, A127
10. Alcalá, J. M., Natta, A., Manara, C. F., et al. 2014, A&A, 561, A2
11. Rigliaco, E., Natta, A., Testi, L., et al. 2012, A&A, 548, A56
12. Espaillat, C., Muzerolle, J., Najita, J., et al. 2014, Protostars and Planets VI, 497
13. Tazzari, M., Testi, L., Natta, A., et al. 2017, A&A, 606, A88
14. Appenzeller, I., & Bertout, C. 2013, A&A, 558, A83
15. Caratti o Garatti, A., Eislöffel, J., Froebrich, D., et al. 2009, ApJ, 502, 579
16. Rigliaco, E., Pascucci, I., Gorti, U., Edwards, S., & Hollenbach, D. 2013, ApJ, 772, 60

Accretion Bursts from Young Stars

Alessio Caratti o Garatti and Jochen Eislöffel

1 Introduction

During the entire formation process, YSOs accrete matter through their disks with decreasing mass accretion rates ($\dot{M}_{acc}$), until the matter reservoir of both envelope and disk is dissipated (e.g. [1, 2]). The accretion process is far from steady, as small variations ($\Delta\dot{M}_{acc} \ll 10$) on timescales from hours to weeks are very common in YSOs. Until few decades ago, outbursting young stars ($\Delta\dot{M}_{acc} > 10$) were considered as a small class of odd objects, not really connected to the large picture of star formation. Historically, these phenomena were observed in the optical and associated to relatively evolved Classical T Tauri stars (CTTs) (age $\geq 10^6$ yr), classified as FUor or EXor objects, depending on the strength and duration of their outbursts. FUors suddenly brighten by 5–6 magnitudes in less than one year [3] and their $\dot{M}_{acc}$ increases by several orders of magnitude (up to $10^{-4}\,M_\odot\,yr^{-1}$), remaining bright for decades. On the other hand, EXors have shorter (months to year) and lower (1–2 orders of magnitude) amplitude outbursts (with $\dot{M}_{acc}$ up to $10^{-7}\,M_\odot\,yr^{-1}$; see [4, 5]). Notably, both sub-classes of objects show very distinct spectral characteristics in the optical and near infrared (NIR). In particular, EXors display emission-line spectra, as seen in CTTs (see e.g., [4]), whereas FUors have absorption-line spectra, which render distinct spectral types (SpTs) as a function of the wavelength (i.e. F or G SpTs can be inferred in the optical, and M SpT in the NIR; see e.g. [5]).

A. Caratti o Garatti (✉)
Astronomy & Astrophysics Section, Dublin Institute for Advanced Studies, Dublin 2, Ireland
e-mail: alessio@cp.dias.ie

J. Eislöffel
Thüringer Landessternwarte Tautenburg, Tautenburg, Germany
e-mail: jochen@tls-tautenburg.de

C. Sauty (ed.), *JET Simulations, Experiments, and Theory*,
Astrophysics and Space Science Proceedings 55,
https://doi.org/10.1007/978-3-030-14128-8_16

On the other hand, in recent years, episodic accretion has gained increasing relevance and the latest picture of YSO evolution suggests that a large portion of the accreted material is gathered during outbursts (see [5] for a recent review). Indeed, most recent observations have corroborated the idea that accretion bursts take place through a broad range of stellar masses, from very low luminosity objects (VeLLOs) [6] to high-mass YSOs [7] and at all stages of star formation [8–11]. Moreover the larger statistics on new eruptive young stars shows that outburst strength, duration, and spectral characteristics are not anymore well separated in two single classes (FUors & EXors), suggesting a continuum rather than a bimodal distribution in the accretion burst typologies. These new eruptive objects have been temporarily called MNORs [12]. A fundamental role in detecting new eruptive YSOs have been played by various IR photometric surveys (as, e.g., VVV or HOPS [11, 13]), which have disclosed the most embedded and young outbursting sources.

As a matter of fact, episodic accretion has challenged the traditional steady-state accretion model, allowing us to solve the long standing luminosity problem in protostars [14] as well as the luminosity spread in young clusters and their age inconsistencies within the HR diagrams [15]. Moreover, this new picture significantly affects our understanding of star and planet formation [16, 17], binary formation [18], and disk chemistry [19].

This paper reviews three outbursting sources with outstanding characteristics, namely V2775 Ori, V960 Mon, and S255IR NIRS 3.

2 V2775 Ori: A Low-Mass Class I FUor

V2775 Ori is one of the less massive ($M_* \sim 0.3\,M_\odot$) and youngest FUors so far detected [9, 20]. The date of V2775 Ori eruption is uncertain (sometime between 2005 and 2007). Pre- and post-outburst observations show an increase in brightness by ~4.6, 4.0, 3.8, and 1.9 mag in the J, H, K_s bands and at 24 μm, respectively, corresponding to L_{bol} and $\dot{M}_{acc}$ increments of ~20–28 $L_\odot$ and $10^{-5}\,M_\odot\,yr^{-1}$, respectively. The pre-outburst spectral energy distribution (SED) indicates a Class I YSO, thus a very young and embedded source, which explains the fact that the object is not detected at optical wavelengths. Its NIR spectrum displays several features typical of a FUor outburst, namely CO, H_2O and VO bands in absorption [9] as well as the He I line at 1.08 μm in absorption, blue-shifted by ~300 km s^{-1} [20], tracing a fast wind in 2011. The line almost disappears in 2015 spectra [21].

3 V960 Mon: A Class II FUor in a Multiple System

V960 Mon (or 2MASS J06593158-0405277) is a low-mass (0.75 $M_\odot$ [22]) young (5×10^5 yr) CTTs undergoing an FU Ori-type outburst since 2014 [23]. VLT/SINFONI high angular resolution AO-assisted observations reveal the presence of an extended disk-like structure around the FUor, a very low-mass companion (SpT = M7 ± 2 with $T_e = 2900 \pm 300$ K) at ~100 AU in projection, and, possibly, a third closer companion (seen as a "bump" in the continuum-subtracted spectral-images) at ~11 AU (Fig. 1). These sources appear to be young, displaying accretion signatures [24], namely Paschen and Brackett lines in emission, in contrast to the bursting source spectrum, which shows the typical FUor features in the NIR, namely Paschen lines in absorption, no Brackett lines, and prominent CO and H_2O bands in absorption. The combination of spectra from spatially unresolved sources (FUor plus other companions) might then explain some strange spectral features observed in some MNors, which display a mixture of FUOr and EXor features.

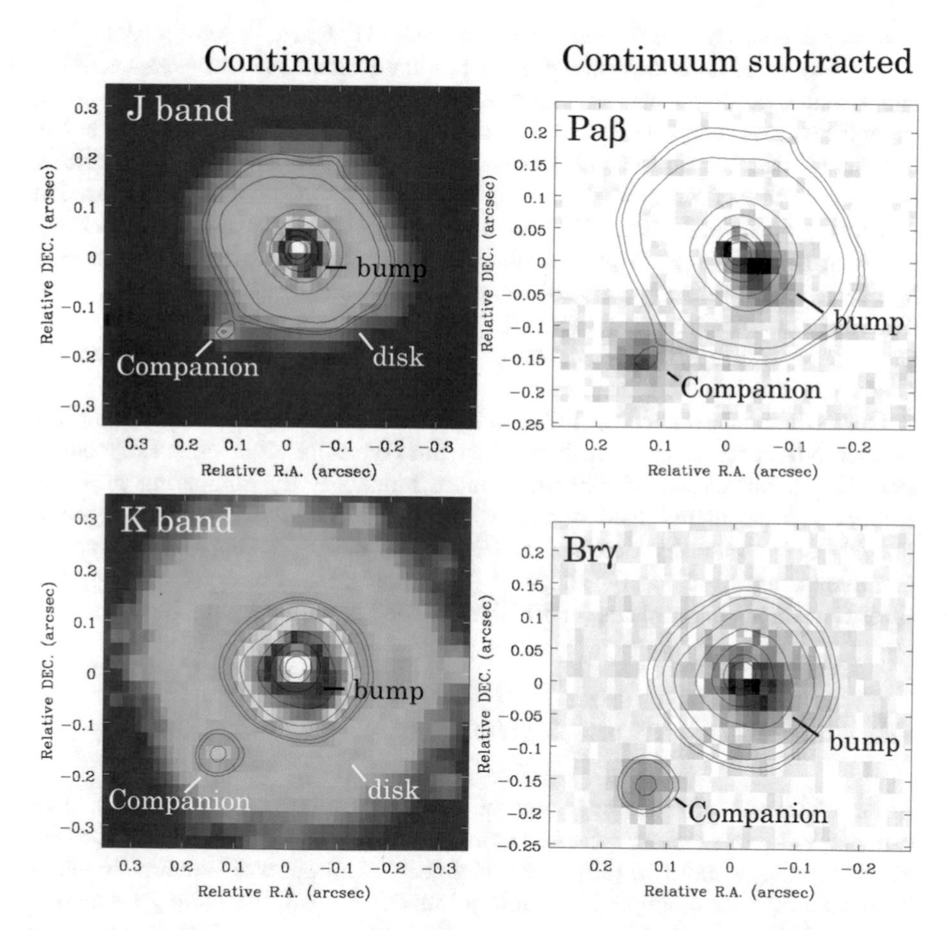

Fig. 1 *Left*: SINFONI J- (upper panel) and K-band (lower panel) images of V960 Mon showing the disk-like feature, the "bump" (possibly a close companion at ~11 au), and the companion positions (B). *Right*: Paβ (upper panel) and Brγ (lower panel) continuum-subtracted grayscale images. (Adapted from [24])

Additional photometric studies from [25] reveal a fast photometric variability of V960 Mon (with a period of about 17 days), suggesting the presence of a close binary with eccentric orbit. Assuming that all these components are physically linked, V960 Mon would then be the first quadruple system ever observed in a FUor.

4 S255IR NIRS3: First Disk-Mediated Outburst in a HMYSO

Evidence of bursts was so far missing in HMYSOs ($M>8\,M_\odot$, $L_{bol}>5\times10^3\,L_\odot$), as the very nature of their formation process is still uncertain. Nevertheless, accretion bursts should develop if massive stars gain mass through disk-mediated accretion, as their low-mass counterparts. Therefore both detection and study of such bursts in HMYSOs are crucial to confirm that the formation mechanism of these objects is similar to that of their lower-mass siblings. As HMYSOs are deeply embedded and represent a small fraction of the YSO population, the chance of observing accretion bursts in HMYSOs is small. During the NIR imaging follow-up of a 6.7 GHz methanol maser flare [26] in the star forming region S255 IR, our team discovered the accretion burst of S255IR NIRS 3 [7, 27], a $\sim$20 $M_\odot$ HMYSO [28]. Our NIR photometry of NIRS 3 revealed an increase in brightness of $\sim$3.8 and 2.5 mag in the H and K bands, respectively.

Our NIR images show the brightening of the central source and its outflow cavities. NIR spectroscopy reveals emission lines typically observed in EXor bursts (see Fig. 2), but orders of magnitude more luminous. By comparing pre- and outburst spectral energy distributions, we were able to derive the burst energetics. The HMYSO luminosity increased by $1.3\times10^5\,L_\odot$ corresponding to a mass accretion rate increment of $5\times10^{-3}\,M_\odot\,yr^{-1}$ [7]. Notably, the accretion burst triggered Class II methanol maser flares (at 6.7 GHz), excited through IR pumping [29].

4.1 Accretion Turns into Ejection

About 14 months after the beginning of the accretion burst, we also detected an exponential increase in the radio flux density from 6 to 45 GHz, exceeding the pre-existing radio jet emission [30]. The flux density at all observed centimetre bands is reproduced with a simple expanding jet model. Indeed, the radio jet emission has been boosted by a sudden increase in the mass loss rate, which is, in turn, a consequence of the accretion burst.

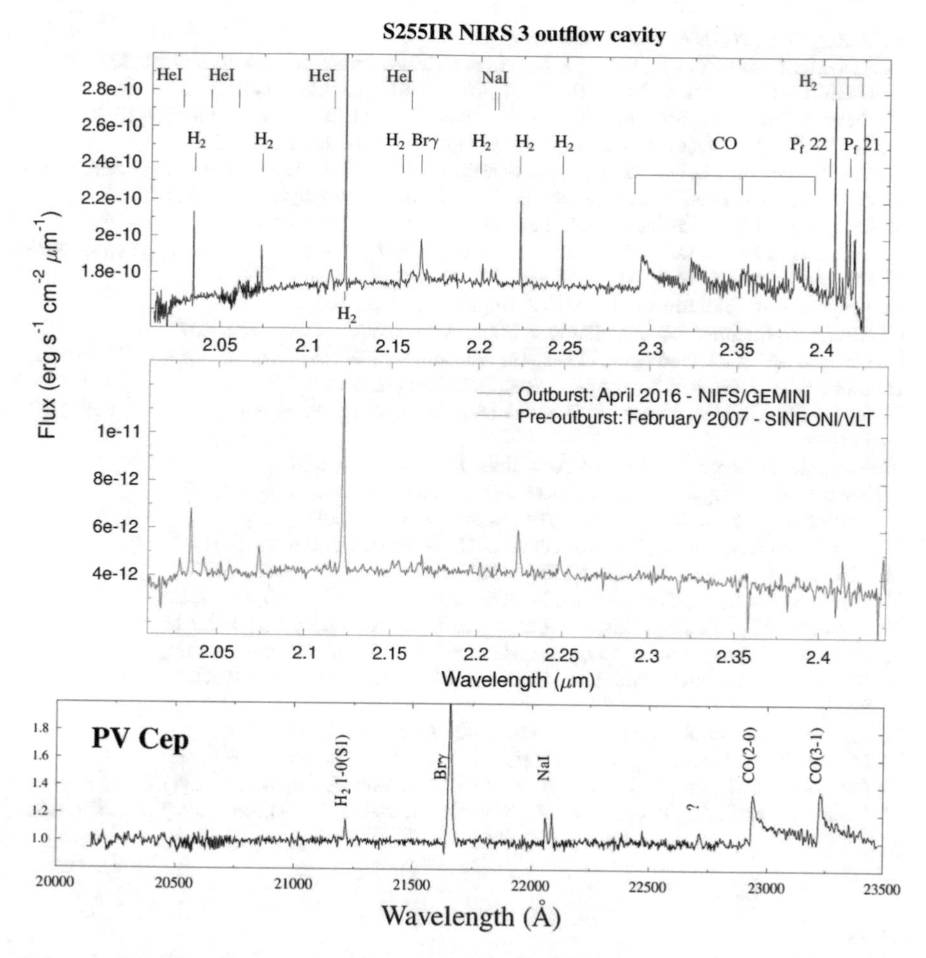

Fig. 2 Pre-outburst (orange) and outburst (black) SINFONI K-band spectra (*middle and top panels*, respectively) of the red-shifted outflow cavity of S255IR NIRS 3. The cavity acts as a mirror allowing to detect the outburst emission. The pre-outburst spectrum only displays H_2 lines in emission, whereas the outburst spectra show emission lines typical of EXor bursts, as e.g. seen in PV Cep (*bottom panel*). (Figure adapted from [7, 31])

Acknowledgements A.C.G. has received funding from the European Research Council (ERC) under the European Union's Horizon 2020 research and innovation programme (grant agreement No. 743029).

References

1. Lada, C. J., & Wilking, B. A. 1984, Astrophys. J., 287, 610
2. Shu, F. H., Adams, F. C., & Lizano, S. 1987, Ann. Rev. Astron. Astrophys, 25, 23
3. Herbig, G. H. 1977, Astrophys. J., 217, 693

4. Herbig, G. H. 2007, Astron. J., 133, 2679
5. Audard, M., Ábrahám, P., Dunham, M. M., et al. 2014, Protostars and Planets VI, 387
6. Hsieh, T.-H., Murillo, N. M., Belloche, A., et al. 2018, Astrophys. J., 854, 15
7. Caratti o Garatti, A., Stecklum, B., Garcia Lopez, R., et al. 2017, Nature Physics, 13, 276
8. Acosta-Pulido, J. A., Kun, M., Ábrahám, P., et al. 2007, Astron. J., 133, 2020
9. Caratti o Garatti, A., Garcia Lopez, R., Scholz, A., et al. 2011, Astron. Astrophys., 526, L1
10. Safron, E. J., Fischer, W. J., Megeath, S. T., et al. 2015, Astrophys. J. Letters, 800, L5
11. Contreras Peña, C., Lucas, P. W., Minniti, D., et al. 2017, Mon. Not. R. Astron. Soc., 465, 3011
12. Contreras Peña, C., Lucas, P. W., Kurtev, R., et al. 2017, Mon. Not. R. Astron. Soc., 465, 3039
13. Furlan, E., Fischer, W. J., Ali, B., et al. 2016, Astrophys. J. Supp., 224, 5
14. Kenyon, S. J., & Hartmann, L. 1995, Astrophys. J. Supp., 101, 117
15. Baraffe, I., Chabrier, G., & Gallardo, J. 2009, Astrophys. J. Letters, 702, L27
16. Dunham, M. M., & Vorobyov, E. I. 2012, Astrophys. J., 747, 52
17. Cieza, L. A., Casassus, S., Tobin, J., et al. 2016, Nature, 535, 258
18. Stamatellos, D., Whitworth, A. P., & Hubber, D. A. 2012, Mon. Not. R. Astron. Soc., 427, 1182
19. Visser, R., & Bergin, E. A. 2012, Astrophys. J. Letters, 754, L18
20. Fischer, W. J., Megeath, S. T., Tobin, J. J., et al. 2012, Astrophys. J., 756, 99
21. Connelley, M. S., & Reipurth, B. 2018, Astrophys. J., 861, 145
22. Kóspál, Á., Ábrahám, P., Moór, A., et al. 2015, Astrophys. J. Letters, 801, L5
23. Maehara, H., Kojima, T., Fujii, M. 2014, The Astronomer's Telegram, 6770
24. Caratti o Garatti, A., Garcia Lopez, R., Ray, T. P., et al. 2015, Astrophys. J. Letters, 806, L4
25. Hackstein, M., Haas, M., Kóspál, Á., et al. 2015, Astron. Astrophys., 582, L12
26. Fujisawa, K., Yonekura, Y., Sugiyama, K., et al. 2015, The Astronomer's Telegram, 8286,
27. Stecklum, B., Caratti o Garatti, A., Cardenas, M. C., et al. 2016, The Astronomer's Telegram, 8732,
28. Zinchenko, I., Liu, S.-Y., Su, Y.-N., et al. 2015, Astrophys. J., 810, 10
29. Moscadelli, L., Sanna, A., Goddi, C., et al. 2017, Astron. Astrophys., 600, L8
30. Cesaroni, R., Moscadelli, L., Neri, R., et al. 2018, Astron. Astrophys., 612, A103
31. Caratti o Garatti, A., Garcia Lopez, R., Weigelt, G., et al. 2013, Astron. Astrophys., 554, A66

The Musca Molecular Cloud: An Interstellar Symphony

Aris Tritsis

1 Introduction

Molecular clouds are often observed to have a "messy" morphology with complex networks of filamentary structures, as a result of turbulence [2]. Despite the general turbulent nature of molecular clouds, a recently discovered type of filamentary structures in their low column density parts, dubbed striations [8], is always observed to be ordered and parallel to the plane-of-sky (POS) component of the magnetic field [1, 5, 8, 10, 12, 13].

The formation mechanism originally proposed for striations was that of streamers along magnetic field lines [1, 12]. In this scenario, material would flow along field lines creating pressure differences due to fluctuations in the streaming speed which would then appear as overdensities in the gas. Alternative formation mechanisms included a Kelvin-Helmholtz instability and fast magnetosonic waves [9].

We performed numerical experiments considering four possible formation mechanisms,

(a) sub-Alfvénic flows along field lines,
(b) super-Alfvénic flows along field lines,
(c) a Kelvin-Helmholtz instability perpendicular to field lines and
(d) fast magnetosonic waves excited naturally by phase mixing with Alfvénén waves.

A. Tritsis (✉)
Research School of Astronomy and Astrophysics, Mount Stromlo Observatory, Canberra, ACT, Australia
e-mail: Aris.Tritis@anu.edu.au

C. Sauty (ed.), *JET Simulations, Experiments, and Theory*, Astrophysics and Space Science Proceedings 55,
https://doi.org/10.1007/978-3-030-14128-8_17

In Sect. 2 we briefly describe the simulations and present our results. In Sect. 3 we explore the predictions of the successful model for the formation of striations.

2 The Physics of Striations

We used the FLASH code [6, 7] to perform 2D and 3D ideal magnetohydrodynamics (MHD) simulations without self-gravity in Cartesian coordinates [14]. For the magnetic field strength and density considered in our simulations, we adopted typical values observed in regions with striations [4].

In the model involving sub-Alfvén ic flows we introduced a constant flow and perturbations in the velocity component parallel to the magnetic field. The perturbations were allowed to be up to 100% the value of the constant flow which was set at ~0.4 times the value of the Alfvén speed. In the super-Alfvén ic flows model we introduced oppositely directed flows along field lines such that their velocity difference was higher than twice the value of the Alfvén speed and the stability condition for the onset of a Kelvin-Helmholtz instability [3] was violated. In the model involving a Kelvin-Helmholtz instability we introduced a sub-Alfvén ic flow perpendicular to field lines which varied along the direction of the magnetic field. Finally, in the MHD waves model, our initial setup involved an Alfvén wave passing through the computational region and density inhomogeneities. In this model, phase mixing of Alfvén waves leads to the excitation of fast magnetosonic waves.

For each of these simulations and for CO (J = 1–0) observations of striations in Taurus molecular cloud we computed the mean contrast between adjacent striations. We found that the mean contrast for the sub-Alfvén ic model was $4 \cdot 10^{-4}$%. For the super-Alfvén ic model the mean contrast was $8 \cdot 10^{-3}$% and for the model involving a Kelvin-Helmholtz instability perpendicular to field lines, the mean contrast between adjacent striations was $4 \cdot 10^{-3}$%. Finally, for the MHD waves model, the mean contrast between adjacent striations was up to ~7%. These values should be compared to a mean contrast of ~25% derived from observations. Only the MHD waves model can account for the observed contrast between adjacent striations.

Additionally, for each formation mechanism tested with numerical simulations we computed the spatial power spectra of cuts perpendicular to the long axis of striations. Results from simulations of each model were then compared to observational results. Only the model involving MHD waves was able to qualitatively reproduce the observed power spectra. The left panel of Fig. 1 shows a CO (J = 1–0) integrated map of striations in Taurus molecular cloud where they were first discovered [8]. A density map from our simulations for the model involving the excitation of fast magnetosonic waves is shown in the right panel.

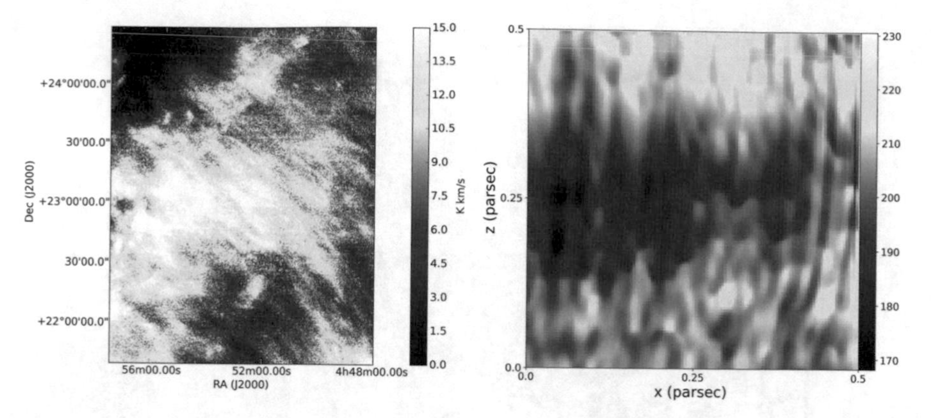

Fig. 1 The left panel shows a CO (J = 0–1) integrated map of striations in Taurus. In the right panel we show a density map from our numerical simulations in which the formation mechanism involves the excitation of fast magnetosonic waves

3 The Song of Musca

Since striations are formed from fast magnetosonic waves, a direct prediction is that when the propagation speed of these waves changes, they could be reflected. Thus, the waves forming striations may get trapped in the presence of boundaries, setting up normal modes.

In order to test this prediction we analysed observations of striations in Musca molecular cloud [15]. A dust emission map of Musca is shown in Fig. 2. Because of its morphology, Musca was considered to be the poster-child of an interstellar filament. We considered cuts perpendicular to the striations inside the green region shown in Fig. 2 and computed their power spectra.

The power spectra from all cuts perpendicular to striations are shown in the left panel of Fig. 3. A distribution of peaks (marked with red dots in the left panel of Fig. 3) from the power spectra is shown in the right panel of Fig. 3. We then tested how analytical relations for the normal modes of possible geometries compare to that distribution. For a rectangular box, the spatial frequencies of normal modes (k_{nm}) are given by:

$$k_{nm} = \sqrt{(\frac{\pi n}{L_x})^2 + (\frac{\pi m}{L_y})^2} \tag{1}$$

where the magnetic field is assumed to be along the z axis, L_x and L_y are the lengths of the box in the directions perpendicular to the magnetic field, and n and m are integers.

In the distribution of peaks, the first two peaks, are marked with red lines. From the value of the wavenumber at the first peak and by setting $(n, m) = (1, 0)$ (i.e. the set of integers that corresponds to the minimum wavenumber) in Eq. 1, we obtain $L_x =$

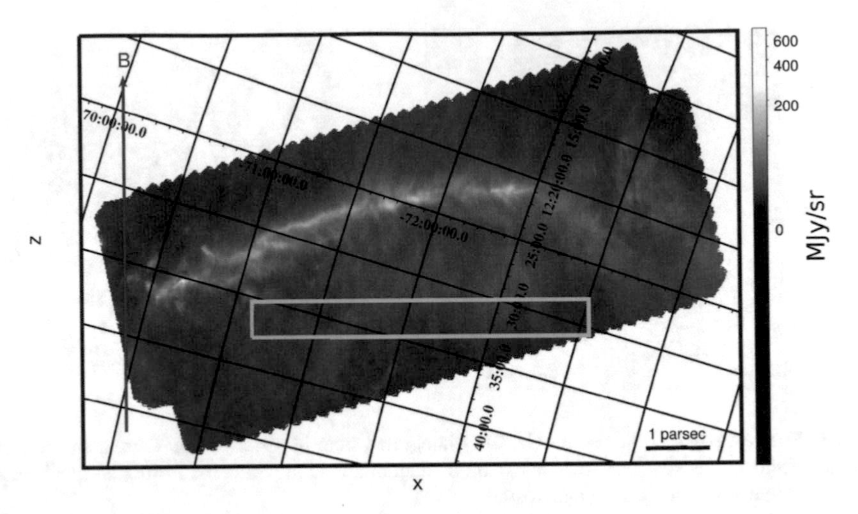

Fig. 2 *Herschel* dust emission map (250 μm) of the Musca molecular cloud. The image shows the dense elongated structure with perpendicularly oriented striations. The blue arrow marks the approximate direction of the magnetic field as probed by polarization measurements [5]. The green region is where we performed our normal mode analysis

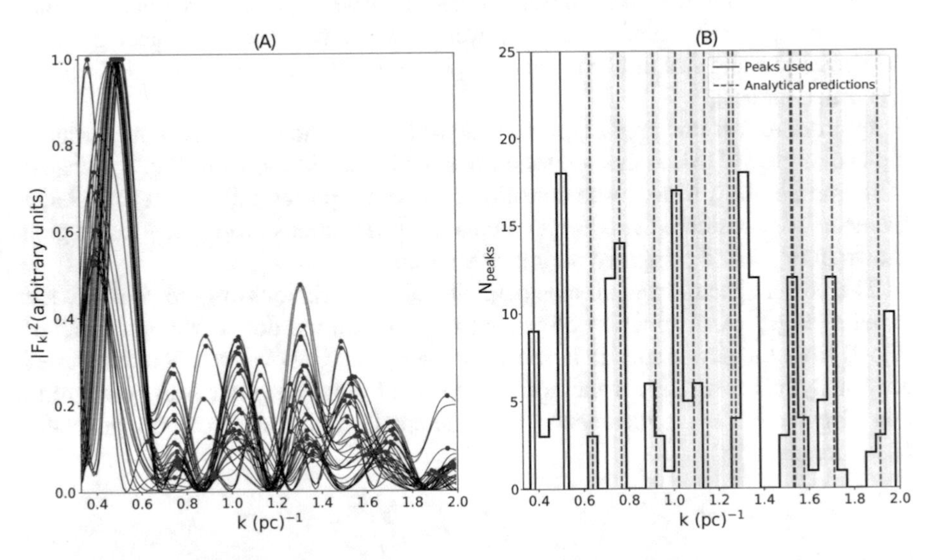

Fig. 3 Left panel: normalized power spectra from all cuts perpendicular to the striations. The red dots show the peaks identified. Right panel: Distribution of peaks of the power spectra at each wavenumber. The two red lines show the two peaks we used to measure the length of the cloud in each direction perpendicular to the magnetic field. The blue dotted lines and shaded regions (1σ uncertainty) show the frequencies predicted analytically from Eq. 1

8.2 ± 0.3 parsecs. With the derived value for L_x and for $(n, m) = (2, 0)$ the predicted peak from Eq. 1 lies at $\sim$0.8 (parsecs)$^{-1}$. This value is much higher than the actual location of the second peak found from observations. Thus the second peak in the right panel of Fig. 3 has to correspond to $(n, m) = (0, 1)$. From the value of the wavenumber at the second peak and $(n, m) = (0, 1)$ we find $L_y = 6.2 \pm 0.2$ parsecs. We insert the values for L_x and L_y into Eq. 1 and we predict the spatial frequencies for integers up to $(n, m) = (3, 3)$. Our results are shown with the blue dotted lines in the left panel of Fig. 3 and are in excellent agreement with observations. The shaded regions around the blue dotted lines are the 1σ uncertainty from error propagation because of uncertainties in the location of the first two peaks used to derive L_x and L_y.

This is the first time that normal modes were discovered in such large scales and the first time the dimensions of a cloud were measured with such low uncertainties. Since the derived sizes of the cloud perpendicular to the magnetic field (L_x and L_y) are comparable, Musca is not a filament as previously considered but is rather a sheet seen edge on. This shape can be naturally explained in the case where the magnetic field is dynamically important in the evolution of the cloud. Magnetic pressure forces are exerted on the direction perpendicular to the magnetic field, while along field lines the cloud can collapse freely under its self-gravity, until thermal pressure becomes important. Thus, in the resulting configuration the ordered magnetic field is parallel to the shortest axis of dense structures (e.g. [11]). Finally, the discovery of normal modes in Musca, being a direct prediction of the model for the formation of striations involving fast magnetosonic waves, also comes as strong supporting evidence for its validity.

However, the existence of normal modes was not the only prediction arising from the formation model of striations developed by [14]. By linearising the equations of ideal MHD and assuming a plane-wave solution for the displacement of the gas, it can be shown that the power spectra of column density and velocity cuts perpendicular to striations should peak at the same wavenumbers [16]. This prediction was recently confirmed using observations of striations in Taurus molecular cloud (see Fig. 4).

Additionally, by considering the ratio of powers between the velocity and column density spectra and combining it with the dispersion relation of fast magnetosonic waves and the expression for the Alfvén speed one can obtain [16]:

$$B_{pos} = \Gamma_n N_{H_2} \sqrt{4\pi\rho} \tag{2}$$

where B_{pos} is the POS component of the magnetic field, Γ_n is the ratio of powers of velocity to column density power spectra at different peaks (k_n), N_{H_2} is the mean column density and ρ is the mean density of the region with striations. Using Eq. 2, [16] found that for the striations in Taurus molecular cloud, the POS component of the magnetic field was $27 \pm 7\,\mu$G in agreement with previous estimates [4]. This method was also successfully applied to elongated structures in HI clouds, dubbed "fibers" [17].

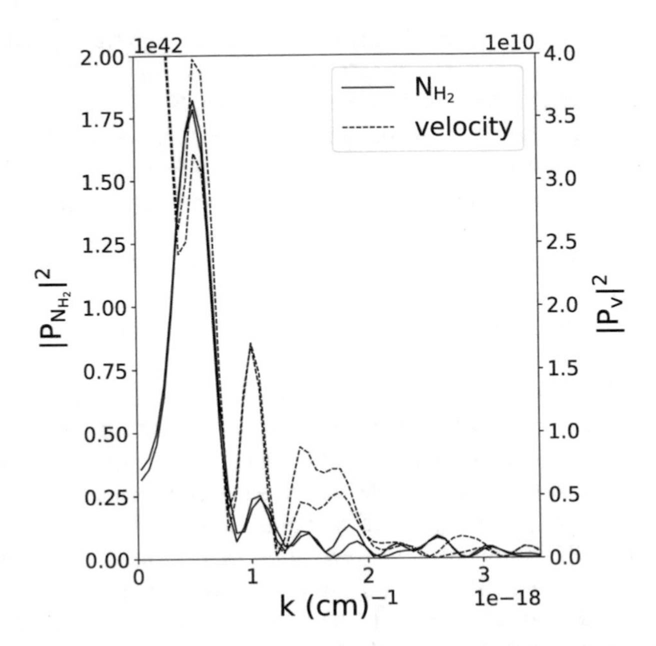

Fig. 4 Column density (solid lines) and velocity (dashed lines) power spectra from cuts perpendicular to the striations in Taurus molecular cloud (left panel of Fig. 1). The power spectra peak at approximately the same wavenumbers

Acknowledgements We thank K. Tassis, C. Federrath, N. Schneider, V. Pavlidou, G. Panopoulou, V. Charmandaris, N. Kylafis, A. Zezas, E. Economou, J. Andrews, S. Williams, P. Sell, D. Blinov, I. Liodakis, T. Mouschovias for comments that helped improve this paper. Usage of the Metropolis HPC Facility at the Crete Center for Quantum Complexity and Nanotechnology of the University of Crete, supported by the European Union Seventh Framework Programme (FP7-REGPOT-2012–2013-1) under grant agreement no. 316165, is acknowledged.

References

1. Alves de Oliveira, C., Schneider, N., Merín, B., et al. Herschel view of the large-scale structure in the Chamaeleon dark clouds. Astron. Astrophys., **568**, A98 (2014)
2. André, P., Di Francesco, J., Ward-Thompson, D., et al. From filamentary networks to dense cores in molecular clouds: toward a new paradigm for star formation. *Protostars and Planets VI*, 27–51 (2014)
3. Chandrasekhar, S. Hydrodynamic and hydromagnetic stability. International Series of Monographs on Physics, Oxford: Clarendon, (1961)
4. Chapman, N. L., Goldsmith, P. F., Pineda, J. L., et al. The Magnetic Field in Taurus Probed by Infrared Polarization. Astrophys. J., **741**, 21 (2011)
5. Cox, N. L. J., Arzoumanian, D., André, P., et al. Filamentary structure and magnetic field orientation in Musca. Astron. Astrophys., **590**, A110 (2016)
6. Dubey, A., Fisher, R., Graziani, C., et al., Challenges of Extreme Computing using the FLASH code. Numerical Modeling of Space Plasma Flows, **385**, 145 (2008)

7. Fryxell, B., Olson, K., Ricker, P., et al. FLASH: An Adaptive Mesh Hydrodynamics Code for Modeling Astrophysical Thermonuclear Flashes. Astrophys. J. Supp., **131**, 273 (2000)
8. Goldsmith, P. F., Heyer, M., Narayanan, G., et al. Large-scale structure of the molecular gas in Taurus revealed by high linear dynamic range spectral line mapping. Astrophys. J., **680**, 428–445 (2008)
9. Heyer, M., Goldsmith, P. F., Yıldız, U. A., et al. Striations in the Taurus molecular cloud: Kelvin-Helmholtz instability or MHD waves? Mon. Not. R. Astron. Soc., **461**, 3918 (2016)
10. Malinen, J., Montier, L., Montillaud, J., et al. Matching dust emission structures and magnetic field in high-latitude cloud L1642: comparing Herschel and Planck maps. Mon. Not. R. Astron. Soc., **460**, 1934–1945 (2016)
11. Mouschovias, T. C. Star formation in magnetic interstellar clouds. I - Interplay between theory and observations. II - Basic theory. NATO ASIC Proc. 210: Physical Processes in Interstellar Clouds, **453** (1987)
12. Palmeirim, P., André, P., Kirk, J., et al. Herschel view of the Taurus B211/3 filament and striations: evidence of filamentary growth? Astron. Astrophys., **550**, A38 (2013)
13. Panopoulou, G. V., Psaradaki, I., & Tassis, K. The magnetic field and dust filaments in the Polaris Flare. Mon. Not. R. Astron. Soc., **462**, 1517–1529 (2016)
14. Tritsis, A., & Tassis, K. Striations in molecular clouds: streamers or MHD waves? Mon. Not. R. Astron. Soc., **462**, 3602 (2016)
15. Tritsis, A., & Tassis, K. Magnetic seismology of interstellar gas clouds: Unveiling a hidden dimension. Science, **360**, 635 (2018)
16. Tritsis, A., Federrath, C., Schneider, N., & Tassis, K. Mon. Not. R. Astron. Soc., (2018a)
17. Tritsis, A., Federrath, C., & Pavlidou, V. arXiv:1810.00231 (2018b)

Laboratory Astrophysics at Extreme Light Infrastructure: Nuclear Physics

Ovidiu Tesileanu, for the ELI-NP team

1 Introduction

The buildings of ELI-NP and a significant part of the equipment are completed [1, 2] in Magurele, Romania, where a large fraction of the Physics research institutes of the country are located.

The High Power Laser System, to provide two beams with the highest laser pulse power in the world, of 10 PW, is already installed and in an advanced testing stage. The system will be commissioned at full power in the first half of 2019.

This facility will break new ground in the peak laser intensities attained worldwide, going beyond 10^{23} W/cm^2 for the 10 PW beams in tight focus. In order to achieve this, a commissioning (ramp-up) period is needed, for which we have an already proposed experimental programme, endorsed by an International Scientific Advisory Board, that will demonstrate the top-notch capabilities.

Among the proposed experiments there are many relevant for astrophysics, but since ELI-NP will be a user-facility, receiving applications for beamtime from any research team worldwide, we will make in the following sections an overview of the beam capabilities and interaction areas available.

O. Tesileanu (✉)
Extreme Light Infrastructure-Nuclear Physics (ELI-NP), "Horia Hulubei" National R&D Institute for Physics and Nuclear Engineering (IFIN-HH), Magurele, Romania
e-mail: ovidiu.tesileanu@eli-np.ro

C. Sauty (ed.), *JET Simulations, Experiments, and Theory*,
Astrophysics and Space Science Proceedings 55,
https://doi.org/10.1007/978-3-030-14128-8_18

2 The Laser Beams

The European distributed research infrastructure ELI aims to push ahead the power of the laser pulses available for experiments beyond the 10 PW mark, implementing in the three centers in Romania, Hungary and the Czech Republic high power lasers based on different technologies.

In Romania, at ELI-NP, the laser system is developed on the Ti:Sapphire solid state technology, for this purpose the largest such crystals, of 200 mm diameter, being grown. 48 pump lasers at 532 nm increase the energy of the stretched laser pulse with center wavelength at 810 nm along the two beam lines. The final amplification stage, featuring the large Ti:Sapph crystals, is pumped by very stable 100 J pump lasers developed specifically for this project. The pulses are then compressed to a duration below 25 fs in vacuumed compressors with large optical gratings. The 10 PW pulses will be delivered at a repetition rate of one pulse per minute.

Responding to the call for research in applied physics, that could transfer benefits back to society in short periods of time, but also to the need of gradual increase in laser pulse power for some experiments, outputs at intermediate powers of 100 TW and 1 PW are available. The pulses have in these cases the same ultra-short duration of 22–25 fs, but higher repetition rates of 10 and 1 Hz, respectively.

After the compressors, the ultrashort pulses will travel in the vacuum pipes of the beam transport system. The 10 PW pulses have 55 cm diameter in full aperture (and the beam transport pipes 800 mm), so the mirrors for the transport system will have dimensions close to one meter. Extremely good quality of the optics, allowing no more that $\lambda/20$ RMS deviations, is also required in order to keep the characteristics of the laser pulse until it reaches the focusing mirror.

The laser beams are transported to five experimental areas that will be described in the next section.

Due to the large distances of propagation of the laser beams from the laser bay to the experimental areas, amounting to 30–60 m, the pointing stability and the divergence parameters of the laser where very strictly controlled.

3 Experimental Highlights

The eight experimental halls of ELI-NP are grouped on a vibration-stabilized 1.5 meter-thick concrete platform together with the bays for the gamma beam system with radiation-protection provisions, high dose rates being produced in the interaction between the intense focused laser pulses and solid or gas targets. In Fig. 1, an overview of the radiation-protected bunkers is shown.

On the other hand, in short-pulse laser-based experiments very intense electromagnetic pulses (EMP) may be produced. In order to protect the equipment from these, several layers of protection have been set – beginning with a minimum 60 dB

Fig. 1 View of the hall of the experimental areas at ELI-NP

damping of the vacuum interaction chambers themselves, continuing with the EMP filters for all cable or pipe ducts and ending with the insulation embedded in the concrete radioprotection walls and the conductive cover on top of the removable ceiling of the experimental areas.

Five of the experimental areas get access to the ultrashort laser pulses, one of them being at the crossroads between the 10 PW laser beams and the high intensity gamma ray beam.

For the main topics of the experimental research at ELI-NP (starting from the more detailed description of the first, commissioning experiments), Technical Design Reports (TDRs) have been devised covering each experimental area.

3.1 Laser-Based Experiments Relevant for Astrophysics

One of the main research aims of ELI-NP is to combine the precise methods of measurement of Nuclear Physics with the novelty of the high power laser experiments. There are two ways in which the ultrashort laser pulses can help performing experiments in the area of "laboratory astrophysics".

The first one is the acceleration of high density bunches of particles up to relativistic energies, over short distances, these bunches being then collided with secondary (and sometimes tertiary) targets producing the nuclear reactions of interest. As an example of nuclear astrophysics experiment, at ELI-NP is foreseen (as presented in the TDR [3]) the acceleration and fission of heavy nuclei (such as ^{232}Th) in a double-layer solid target, and then obtaining very neutron-rich isotopes from the fusion processes of the fission fragments in a secondary double-layer target. This is a typical experiment which could not be done at classical accelerators due to the very low rates involved, but fusion products could become detectable at the very high densities in laser-based acceleration. The very neutron rich isotopes are a still unknown area of the chart of nuclides, but important for the nucleosynthesis of heavy elements in the Universe. The experimental area E1 is foreseen for this type of experiments, featuring a big 24 m^3 vacuum interaction chamber that can accommodate two short-focus off-axis parabola mirrors for the focusing of the 10 PW laser beams.

A second way of performing astrophysically-relevant experiments is to obtain the extreme plasma conditions encountered in astrophysical scenarios such as heavy stars or supernova explosions. One could then get information on the thermally-populated nuclear states in these conditions, and study effects such as electron screening, that may impact on the real reaction rates.

Another proposal in the ELI TDRs foresees the use of both the laser pulses and the gamma radiation pulses in a same nucleosynthesis experiment – the laser pulses being used for the creation of the extreme conditions and the population of excited (isomer) states in some nuclei of interest, and the gamma photons exciting nuclear transitions departing from the isomer states [4]. The experimental setup for this experiment (see Fig. 2) has been designed and comprises large volume vacuum enclosures in Aluminium alloy (to minimize nuclear activation over time) in the E7 experimental area.

The two-beam configuration available in each experimental area (for 100 TW, 1 PW and 10 PW power) allows for an extremely broad range of geometries for interaction, making possible experiments of simulation in laboratory of astrophysical jets. Studies of the scaling laws for MHD systems [5] are promising. In this respect, there have been experiments performed at existing facilities [6–8] and simulations [9], which are a natural evolution after the experiments of plasma jets at Imperial College London [10] initiated in the framework of the JETSET project [11]. The possibility to employ both ultra-short and longer laser pulses at ELI-NP is investigated.

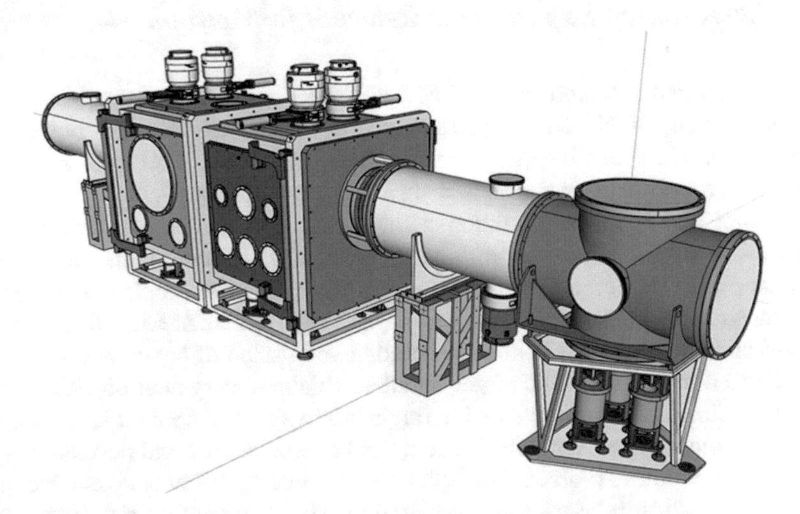

Fig. 2 Vacuum interaction chamber for experiments with combined laser and gamma radiation pulses

3.2 Gamma-Based Experiments of Nuclear Astrophysics

The gamma radiation beam at ELI-NP will have unparalleled characteristics – the possibility to continuously tune the photon energy in a broad range, from 0.2 up to 19.5 MeV, with a very narrow bandwidth of less than 0.5% and high brilliance, of more than 10^3 ph/eV/s. A broad range of detectors are in advanced stages of development at ELI-NP for taking advantage of these characteristics in new experiments [12].

The narrow bandwidth of the gamma beam and good collimation (due to the method of producing the gamma photons, namely Compton backscattering of a pico-second laser pulse on relativistically accelerated electrons) allows a great increase in the signal-to-noise ratio for the study of nuclear reactions of interest. Therefore, experiments of measuring reaction rates relevant for the $s - process$ branching points in nucleosynthesis, as well as photodisintegrations important for the $r - process$, have been already planned [13]. Neutron and gamma detector arrays are constructed at ELI-NP for these studies, for rates and time-of-flight measurements: a 30 ^{3}He counters array, a hybrid gamma-neutron detector array of 3 m diameter and a segmented clover detector array.

Charged particle detectors, both for gas targets (an electronic-readout Time Projection Chamber [14]) and solid targets (a Silicon-Strip Detectors array [15]) with high resolution were designed and are now under development. They will allow the more precise study of reactions of great importance for astrophysics, such as, for example, the $^{12}C(\alpha, \gamma)^{16}O$ reaction through inverse kinematics [16].

4 Conclusions

The ELI-NP project in Romania implements now the base equipment of the experimental areas and is close to commissioning the high power laser system. A first call for experiment proposals to the international community is due in 2019.

Thanks to the various multi-beam configurations that will be available, a wide range of astrophysically-relevant experiments will be possible. Applications in the "laboratory astrophysics" of stellar and extragalactic jets are thought to be possible and will be a great test-bench for the numerical simulations and for the theories on the formation and evolution of these fascinating structures in the Universe.

Acknowledgements Work has been supported by the Extreme Light Infrastructure Nuclear Physics (ELI-NP) Phase II, a project co-financed by the Romanian Government and the European Union through the European Regional Development Fund and the Competitiveness Operational Programme (1/07.07.2016, COP, ID 1334).

References

1. Balabanski, D., Popescu, R., Stutman, D., Tanaka, K., Tesileanu, O., Ur, C.A., Ursescu, D., Zamfir, N.V.: New light in nuclear physics: The extreme light infrastructure. EPL **117**, 28001 (2017)
2. Gales, S., Tanaka, K.A., Balabanski, D., Negoita, F., Stutman, D., et al.: The Extreme Light Infrastructure?Nuclear Physics (ELI-NP) facility: new horizons in physics with 10 PW ultra-intense lasers and 20 MeV brilliant gamma beams. Rep. Prog. Phys. **81**, 094301 (2018)
3. Negoita, F., Roth, M., Thirolf, P.G., Tudisco, S., Hannachi, F., et al.: Laser driven nuclear physics at ELI-NP. Romanian Reports in Physics **68**, Supplement, S37 (2016)
4. Homma, K., Tesileanu, O., D'Alessi, L., Hasebe, T., Ilderton, A., et al.: Combined laser gamma experiments at ELI-NP. Romanian Reports in Physics **68**, Supplement, S233 (2016)
5. Ryutov, D.D., Drake, R.P., and Remington, B.A.: Criteria for Scaled Laboratory Simulations of Astrophysical MHD Phenomena. ApJ. Suppl. **127**, 465 (2000)
6. Budil, K.S., Gold, D.M., Estabrook, K.G., Remington, B.A., Kane, J., et al.: Development of a radiative-hydrodynamics testbed using the Petawatt laser facility. ApJSS, **127**, 261 (2000)
7. Albertazzi, B., Ciardi, A., Nakatsutsumi, M., Vinci, T., Béard, J., et al.: Laboratory formation of a scaled protostellar jet by coaligned poloidal magnetic field. Science **346**, 325 (2014)
8. Albertazzi, B., Chen, S.N., Antici, P., Böker, J., Borghesi, M., et al.: Dynamics and structure of self-generated magnetics fields on solids following high contrast, high intensity laser irradiation. Physics of Plasmas **22**, 123108 (2015);
9. Ciardi, A., Vinci, T., Fuchs, J., Albertazzi, B., Riconda, C., Pé pin, and Portugall, O.:Astrophysics of Magnetically Collimated Jets Generated from Laser-Produced Plasmas. Phys. Rev. Lett. **110**, 025002 (2013)
10. Burdiak, G.C., Lebedev, S.V., Drake, R.P., Harvey-Thompson, A.J., Swadling, G.F., et al.: The production and evolution of multiple converging radiative shock waves in gas-filled cylindrical liner z-pinch experiments. High Energy Density Physics **9**, 52 (2013)
11. Ciardi, A., Lebedev, S.V., Frank, A., Blackman, E.G., Chittenden, J.P., et al.: The evolution of magnetic tower jets in the laboratory. Physics of Plasmas **14**, 056501 (2007)
12. Filipescu, D., Balabanski, D., Camera, F., Gheorghe, I., Ghita, D., et al.: Future Prospects of Nuclear Reactions Induced by Gamma-Ray Beams at ELI-NP. Physics of Particles and Nuclei **48**, 134 (2017)
13. Camera, F., Utsunomiya, H., Varlamov, V., Filipescu, D., Baran, V., et al.: Gamma above the neutron threshold experiments at ELI-NP. Romanian Reports in Physics **68**, Supplement, S539 (2016)
14. Cwiok, M., Bieda, M., Bihalowicz, J.S., Dominik, W., Janas, Z., et al.: A TPC Detector for studying photo-nuclear reactions at astrophysical energies with gamma-ray beams at ELI?NP. Acta Phys. Pol. B **49**, 509–514 (2018)
15. La Cognata, M., Anzalone, A., Balabanski, D., Chesnevskaya, S., et al.: Gamma ray beams for Nuclear Astrophysics: first results of tests and simulations of the ELISSA array. Journal of Instrumentation **12**, C03079 (2017)
16. Tesileanu, O., Gai, M., Anzalone, A., Balan, C., Bihalowicz, J.S., et al.: Charged particle detection at ELI-NP. Romanian Reports in Physics **68**, Supplement, S699 (2016)

Part IV
Future Projects

GIARPS/GRAVITY Survey: Broad-Band 0.44–2.4 Micron High-Resolution Spectra of T-Tauri and Herbig AeBe Stars – Combining High Spatial and High Spectral Resolution Data to Unveil the Inner Disc Physics

F. Massi, A. Caratti o Garatti, R. Garcia Lopez, M. Benisty, J. Brand, W. Brandner, S. Casu, D. Coffey, C. Dougados, A. Giannetti, L. Labadie, S. Leurini, L. Moscadelli, A. Natta, M. Pedani, K. Perraut, T. Ray, A. Sanna, and N. Sanna

F. Massi (✉) · L. Moscadelli · A. Natta · N. Sanna
INAF-Osservatorio Astrofisico di Arcetri, Firenze, Italy
e-mail: fmassi@arcetri.astro.it

A. Caratti o Garatti · R. Garcia Lopez · T. Ray
Astronomy & Astrophysics Section, Dublin Institute for Advanced Studies, Dublin 2, Ireland
e-mail: alessio@cp.dias.ie

M. Benisty · C. Dougados · K. Perraut
Université Grenoble Alpes, CNRS, IPAG, Grenoble, France

J. Brand · A. Giannetti
Italian ARC, INAF-IRA, Bologna, Italy

W. Brandner
Max Planck Institute für Astronomy, Heidelberg, Germany

S. Casu · S. Leurini
INAF-Osservatorio Astronomico di Cagliari, Selargius (CA), Italy

D. Coffey
School of Physics, University College Dublin, Belfield, Dublin 4, Ireland

L. Labadie
I. Physikalisches Institut, Universität zu Köln, Köln, Germany

M. Pedani
INAF – Fundación Galileo Galilei, Breña Baja, Spain

A. Sanna
Max-Planck-Institut für Radioastronomie, Bonn, Germany

C. Sauty (ed.), *JET Simulations, Experiments, and Theory*, Astrophysics and Space Science Proceedings 55,
https://doi.org/10.1007/978-3-030-14128-8_19

1 Circumstellar Processes in Young Stellar Objects

Disc structures are ubiquitous in astrophysics. In star formation, they are instrumental in mediating accretion of matter onto new-born stars as well as in ejecting matter through collimated outflows. A widely accepted paradigm is that stars form by gathering matter falling from the disc onto the central protostar. Unfortunately, the early phases of this process are difficult to observe, due to heavy obscuration from the parental gas/dust core and lack of spatial resolution. On the other hand, the subsequent formation stage of pre-main sequence stars (PMSs) is easier to study as the sources are less embedded. In addition, PMSs still exhibit a significant accretion rate (e.g. $\sim 10^{-8}\ M_{\odot}\ \mathrm{yr}^{-1}$ in Classical T-Tauri stars – CTTSs). Typically, PMSs with discs also show collimated bipolar jets, through which $\sim$10% of infalling matter is ejected.

A shared view is that in CTTSs, i.e. low-mass ($M < 2\ M_{\odot}$) PMSs with a circumstellar disc, an intense stellar magnetosphere truncates the inner disc at a few stellar radii and the accreting gas falls onto the stars in funnels along field lines (magnetospheric accretion scenario; for details see review by [9]). However, the more massive class of PMSs, the so-called Herbig Ae/Be (HAeBes) stars ($M \sim 2 - 10\ M_{\odot}$) do not seem to host strong magnetic fields [2, 14]. This suggests that the accretion scenario might change with stellar mass and, at the high-mass end of HAeBes, it might be replaced by untruncated discs impinging on the stellar surface.

Another major issue in disc evolution is how matter in the disc loses angular momentum in order to accrete onto the stellar surface. A few mechanisms have been proposed, namely Magneto Rotational Instability, Hall effect, Gravitational Instability, Disc Winds (e.g., [10–13]), but none of these appears to be conclusive (e.g. [9]). Recently, [8] has proposed that the early accretion stage may leave a massive inner disc, which subsequently feeds a residual accretion due to low viscosity levels, so there is no requirement to involve matter in the whole disc as usually believed. In principle, the collimated outflow can carry angular momentum away, but the collimating driver itself is still unclear. Models include magneto-centrifugally driven winds from the accretion disc, along field lines from (i) the stellar surface, or (ii) the region of interaction between the stellar magnetic field and the disc (i.e. X-wind), or (iii) the disc surface (i.e. disc-wind). Determining the mechanism responsible requires a knowledge of size, width, and geometry of launching regions.

Both accretion and ejection occur in the innermost parts (few AU) of a circumstellar disc. As these processes are intimately linked to the global evolution of the disc, they determine the evolution of both stars and planets. The innermost few AU of circumstellar discs are thus crucial in testing the different scenarios which lead to a comprehensive physical picture of the system formation. Furthermore, it is also critical to understand how this picture changes with the stellar mass to gain a full grasp of the first stages of stellar and planetary system evolution.

2 Aims of the GIARPS/GRAVITY Survey

Ideally, the following questions should be addressed via observations, to gain an in-depth view of star-disc interactions:

- What accretion mechanisms operate in stars of different masses?
- What are the physical conditions and accretion rates of the accreting gas streams?
- The predictions of which models best fit spatially resolved observations of jet launching regions?

Clearly, we need to probe the physical conditions and kinematics of gas inside ~1 AU to deal with these issues and test the different scenarios. Thus, the relevant kinematics and physical gas components can only be disentangled using a combination of high spatial and high spectral resolution observations. The recent availability of sensitive near-infrared interferometers (GRAVITY at the VLTI) and optical/near-infrared broad-band spectrometers (GIARPS at the TNG) is opening up unprecedented high-resolution observational window, which will facilitate a major step forward in our understanding of accretion processes in YSOs.

GIARPS combines the high-spectral resolution capability of the spectrometers HARPS-N in the optical and GIANO in the near-infrared [5], providing high-spectral resolution ($R \sim 112{,}000$ in the optical to $R \sim 48{,}000$ in the near-infrared) on a band ranging from 0.44 to 2.4 μm. It has been available at the Telescopio Nazionale Galileo (Canary Islands) since 2016. It will be able to kinematically disentangle the various gas components from the stellar photosphere (HARPS-N) to the inner disc (GIANO) and, thanks to the large number of lines *simultaneously* detected, to trace the physical conditions in the emitting gas (Brackett and Paschen HI, and HeI lines, CO overtone bands). Unfortunately, a few critical parameters, such as inner dusty disc size and inclination, as well as size of the line emitting regions, would remain degenerate and can only be determined by high-spatial resolution observations.

However, even for the nearest young stars (~150 pc), 1 AU translates into ~7 mas. Currently, such high spatial resolution can only be obtained with optical/IR interferometry at the ESO/VLTI, where the latest generation beam combiner GRAVITY [7] allows a major improvement in sensitivity and achieves an angular resolution of ~1 mas. On the other hand, GRAVITY can only operate in low and medium spectral resolution (up to $R \sim 4000$) in the K band (1.95–2.5 μm). Furthermore, GRAVITY can only either provide a snapshot view of a system with a limited number of baselines, or produce an image of the stellar system by collecting observations over a larger range of baselines which may take days or months. Therefore, target variability may bias the interferometric information. Meanwhile, GIARPS can complement the GRAVITY spectral data, helping to remove the degeneracy, So, the high-spatial and spectral resolutions achieved by the two instruments are complementary.

Spectral variability is one further powerful tool that allows access to the 3D structure of the inner disc and the circumstellar environment, to better understand the accretion/ejection processes. PMSs of all masses exhibit accretion variability on all timescales, from days to years [1, 9]. Periodic variations of spectrally resolved line profiles have been successfully used to infer the spatial distribution of accreting gas around T-Tauri stars [3]. Variations related to the morphology of the accreting gas are expected to be correlated with the stellar rotation period (few weeks), whereas variations on longer scales (months, years) are related to larger-scale events, e.g. gravitational instabilities in the disc or interactions with closer companions (see e.g. [4]).

3 Programme Status

The GIARPS/GRAVITY survey takes advantage of GTO time with GRAVITY (PI: R. Garcia Lopez) already allocated to a consortium of European Institutes for high-spatial resolution observations of young stellar objects. The consortium has been awarded 20 nights on the UT and 120 nights on the AT telescopes at the VLTI. The programme is in progress and aims to observe a sample of more than 100 YSOs of different mass, age, and accretion rate. The first results are discussed in [6].

A collaboration between the GRAVITY GTO programme and a team of Italian astronomers stemmed from the availability of GIARPS at Italy's Telescopio Nazionale Galileo. The GIARPS/GRAVITY survey will yield high-resolution spectra of both T-Tauri stars and HAeBes from the list of GRAVITY targets. A pilot programme (PI: F. Massi) of 3 nights was carried out in December 2017, during the Italian semester AOT37, and a first sample of 17 targets were observed. Another 7 nights have already been allocated in two semesters (20018B and 2019A) of international CCI-ITP time (PI: A. Caratti o Garatti). Finally, 8 h have been allocated to a pilot programme for studying spectral variability of PMSs (PI: F. Massi) in the Italian semester AOT38 (December 2018).

Figure 1 shows a sample of GIARPS spectra, still uncalibrated, from four Herbig Be stars (spectral type from B0 to B9) obtained during run AOT37. The optical band displays photospheric hydrogen absorption lines (Balmer series) in two cases (GU CMa and HD37806), some lines partially filled with emission. The other two stars (MWC 137 and Z CMa) exhibit Balmer lines in emission. Due to the low S/N, these spectra will be resampled to a lower spectral resolution. Hydrogen Hα is also shown, always in emission. As for the near-infrared band, Fig. 1 shows hydrogen Paβ, Paγ, and Br16 lines. All these lines are in emission. Interestingly, the Paschen and Brackett profiles for MWC 137 and GU CMa are clearly double-peaked, a signature of a circumstellar disc.

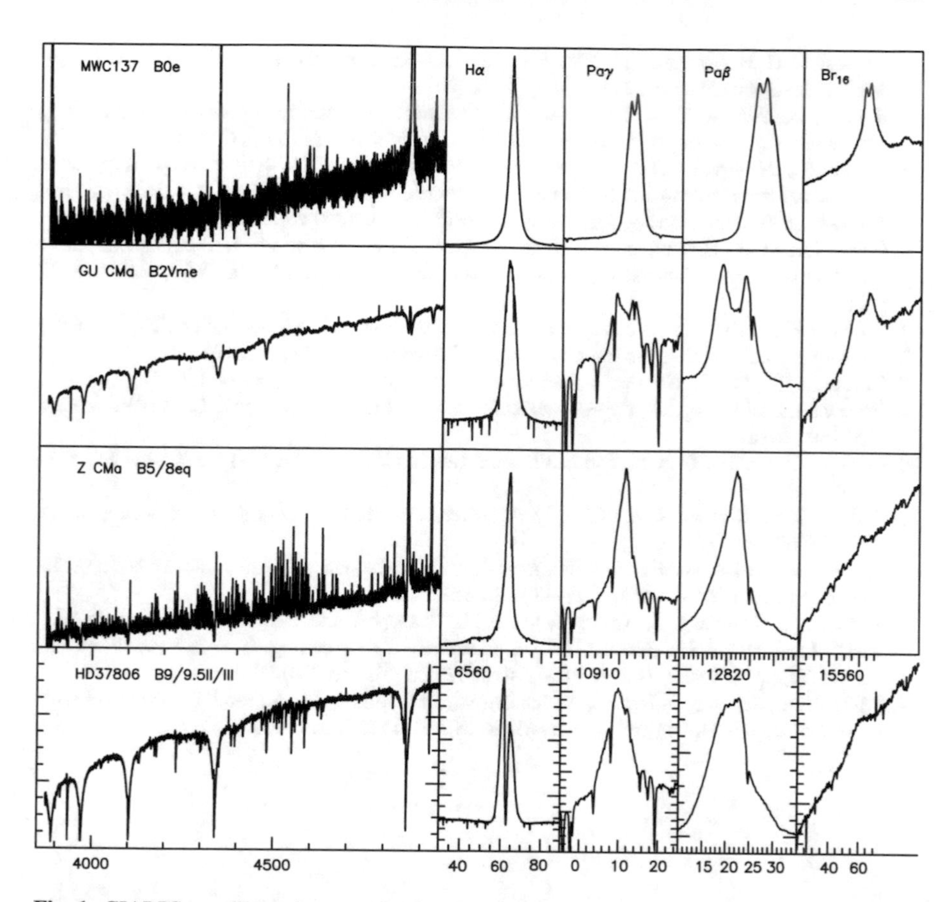

Fig. 1 GIARPS uncalibrated spectra for four Be stars, from top to bottom: MWC 137, GU CMa, Z CMa, and HD37806. The boxes display, from the left: a fraction of the optical band (wavelengths in Å on the bottom axis), and the optical/near-infrared hydrogen lines Hα, Paγ, Paβ, and Br16. Central wavelength, in Å, are indicated on the top left corner of the boxes in the bottom row. The velocity intervals of the near-infrared spectra are: 2700 km s^{-1} (Hα), 800 km s^{-1} (Paγ), 470 km s^{-1} (Paβ), and 1500 km s^{-1} (Br16)

Acknowledgements Based on observations made with the Italian Telescopio Nazionale Galileo (TNG) operated on the island of La Palma by the Fundación Galileo Galilei of the INAF (Istituto Nazionale di Astrofisica) at the Spanish Observatorio del Roque de los Muchachos of the Instituto de Astrofisica de Canarias.

References

1. Aarnio, A.N., Monnier, J.D., Harries, T.J., et al.: High-cadence, High-resolution Spectroscopic Observations of Herbig Stars HD 98922 and V1295 Aquila. ApJ **848**, 18, 34 pp. (2017)
2. Alecian, E., Wade, G. A., Catala, C., et al.: A high-resolution spectropolarimetric survey of Herbig Ae/Be stars – II. Rotation. MNRAS **429**, 1027–1038 (2013)

3. Alencar, S. H. P., Bouvier, J., Walter, F. M., et al.: Accretion dynamics in the classical T Tauri star V2129 Ophiuchi. A&A **541**, A116, 14 pp. (2012)
4. Benisty, M., Perraut, K., Mourard, D., et al.: Enhanced Hα activity at periastron in the young and massive spectroscopic binary HD 200775. A&A **555**, A113, 10 pp. (2013)
5. Claudi, R., Benatti, S., Carleo, I., et al.: GIARPS: the unique VIS-NIR high precision radial velocity facility in this world. In: Evans, C.J., Simard, L., Takami, H. (eds.) Ground-based and Airborne Instrumentation for Astronomy VI, SPIE Proceedings vol. 9908 (2016)
6. Garcia Lopez, R., Perraut, K., Caratti O Garatti, A., et al.: The wind and the magnetospheric accretion onto the T Tauri star S Coronae Australis at sub-au resolution. A&A **608**, A78, 12 pp. (2017)
7. GRAVITY collaboration, First light for GRAVITY: Phase referencing optical interferometry for the Very Large Telescope Interferometer. A&A **602**, A94, 23 pp. (2017)
8. Hartmann, L., Bae, J.: How do T Tauri stars accrete?. MNRAS **474**, 88–94 (2018)
9. Hartmann, L., Herczeg, G., Calvet, N.: Accretion onto Pre-Main-Sequence Stars. ARA&A **54**, 135–180 (2016)
10. Kratter, K., Lodato, G.: Gravitational Instabilities in Circumstellar Disks. ARA&A **54**, 271–311 (2016)
11. Pudritz, R.E., Norman, C.A.: Centrifugally driven winds from contracting molecular disks. ApJ **274**, 677–697 (1983)
12. Simon, J.B., Lesur, G., Kunz, M.W., Armitage, P.J.: Magnetically driven accretion in protoplanetary discs. MNRAS **454**, 1117–1131 (2015)
13. Turner, N.J., Fromang, S., Gammie, C., et al.: Transport and Accretion in Planet-Forming Disks. In: Beuther, H., Klessen, R.S., Dullemond, C.P., Henning, T. (eds.) Protostars and Planets VI, pp. 411–432, University of Arizona Press, Tucson (2014)
14. Wade, G.A., Bagnulo, S.,Drouin, A., Landstreet, J.D., Monin, D.: A search for strong, ordered magnetic fields in Herbig Ae/Be stars. MNRAS **376**, 1145–1161 (2007)

Observational Constraints on the Conservation of Momentum and Energy in Jet-Driven Molecular Outflows

Odysseas Dionatos

1 Introduction

Protostellar ejecta are considered essential mechanisms for removing the excess angular momentum from a protostellar system, allowing further accretion and therefore growth of the nascent star. Jets are well-collimated structures, commonly associated with atomic ejecta from evolved protostars observed in the optical and near-infrared wavelengths, while outflows, typically associated with embedded sources, are traced in molecular lines forming wider, less collimated structures. While jets and outflows appear to coexist to a greater or lesser extent in all phases of star formation [12, 14], it is not yet clear how the two phenomena are linked to each other, or what is their relative influence on the star-formation process over time. Jet entrainment mechanisms can to a certain extent explain the observed morphological characteristics of outflows. To this end, an important theoretical assumption is that the momentum between highly collimated jets and outflows is conserved so that the forward motion of a dissipative bowshock can create the typical outflow-lobe structures observed. In an alternative scenario where the energy is conserved and adiabatically released from the jet to the ambient medium, the observed structures will look like bubbles driven by high-pressure gas in the case of a homogeneous medium [11]. However given the density gradient of protostellar envelopes, the shape of energy-conserved outflows could in principle reproduce outflow lobe-like structures. In this context we here report on the results of the largest comparison between the dynamical and kinematical properties of jets and outflows, based on homogeneous datasets and treated with consistent methods. The sample consists

O. Dionatos (✉)
Institute for Astronomy (IfA), University of Vienna, Vienna, Austria
e-mail: odysseas.dionatos@univie.ac.at

C. Sauty (ed.), *JET Simulations, Experiments, and Theory*,
Astrophysics and Space Science Proceedings 55,
https://doi.org/10.1007/978-3-030-14128-8_20

of seven protostars and their corresponding ejecta, located in the NGC 1333 star-forming region.

2 Deriving the Kinematical and Dynamical Properties of Atomic Jets and Molecular Outflows in NGC 1333

Atomic jets were recovered from observations of the $^3P_1 - ^3P_2$ [OI] line centered at 63.18 μm with the Herschel Space Observatory [8]. While the typical spectral resolution reached by Herschel/PACS does not exceed 100 km s^{-1}, employing a detailed study of the [OI] line centroids through Gaussian fits we reach resolutions of 5 km s^{-1}. This is mainly due to the strength of the [OI] which allows a very accurate determination of the line centroids. The spatial distribution of the [OI] line centroids is presented in the left panel of Fig. 1, showing fast atomic jets associated with embedded (Class 0 and I) protostars in NGC 1333. The degree of

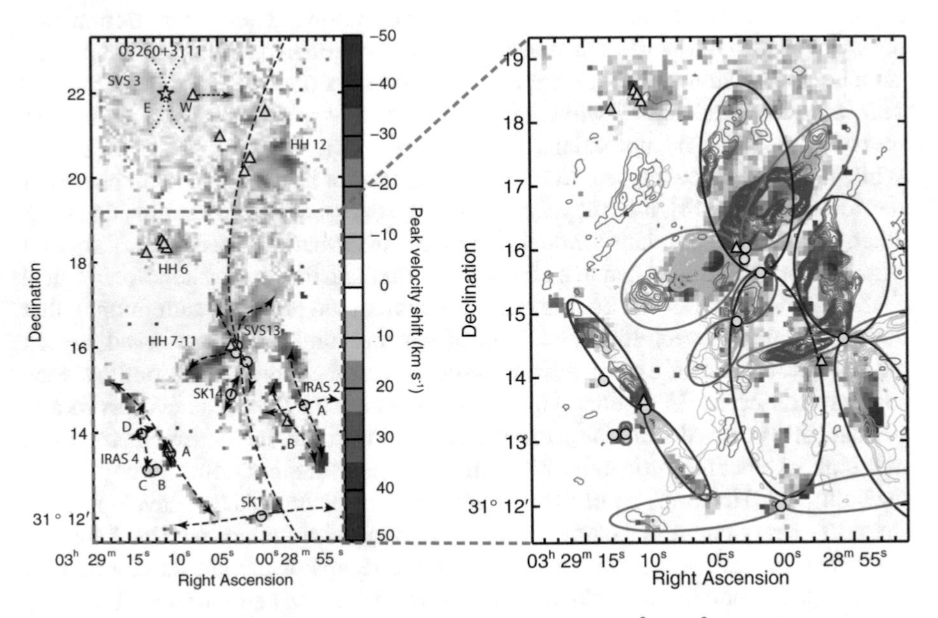

Fig. 1 (*Left:*) Maps displaying the velocity shift observed in the $^3P_1 - ^3P_2$ [OI] line centroids with respect to its rest velocity (from [8]). Color-coding for each pixel corresponds to blue- and redshifted velocities, as indicated on the corresponding bar on the right side of the plot. Atomic jets are marked as dashed arrows and embedded (Class 0 and Class I) sources as open circles and triangles, respectively. The dotted X in the north shows the borderline between the low-velocity blue- and redshifted gas tracing the photodissociative wind from IRAS 03260 + 3111(E) (the position of the source is marked with a star). (*Right:*) The outflow structures as recovered by the [OI] centroid analysis, in comparison with CO interferometric map of [13]. Color levels are the same as in the left panel and the frame now focuses on the southern star-forming complex of NGC 1333. Colored ellipses delineate the outflow lobes attributed to embedded protostars [13] and also employed in our analysis for the consistent description of the atomic jet dimensions

detail recovered by our analysis of the Herschel maps is directly comparable to CARMA interferometric observations of CO 1–0 [13], shown in the right panel of Fig. 1. Both sets of observations recover in most of the cases the same structures, which indicate a strong causal relation between atomic jets and large-scale outflows.

Outflow energy and momenta were calculated employing ^{12}CO, ^{13}CO and C^{18}O maps resulting from combined observations of single-dish and interferometric data [13]. Outflow mass was estimated employing the methods described in [2, 4, 16] employing both ^{12}CO, ^{13}CO to account for opacity effects in the dense regions and low signal-to-noise in more diffuse parts of the cloud (see [13] for a detailed discussion). Then momentum and energies were then calculated using the radial velocity information from the CO lines themselves.

Correspondingly energies and momenta for the [OI] jets were calculated adopting the same geometrical areas used for calculating the outflow properties in [13] (see also Fig. 1). Jet mass was calculated using the method described in [6], while radial velocity information was directly adopted from the [OI] line profile. Mass flux for the atomic jet was calculated however adopting a dynamical timescale that was estimated from the projected dimensions of the jets and proper motion observations reported in [15]. Alternatively, the mass flux of outflows were directly estimated from the relation connecting the mass-loss rate and the [OI] luminosity [9]. A detailed description of the data and the derivation of the kinematical and dynamical properties of the atomic jets can be found in [8].

3 Conservation of Momentum and Energy Between Jets and Outflows

The CO and [OI] momenta appear to closely correlate, as shown in Fig. 2, however the momentum carried along the [OI] line represents only a fraction of $\sim$1% compared to the momentum corresponding to the CO emission. On the other hand, the energy carried by the atomic gas corresponds to 50–100% of the energy measured in the molecular component. In the right panel of Fig. 2, we display the correlation of the [OI] and CO emission with the bolometric luminosity of the driving sources. We note that the star symbols correspond to the east-west directed outflow of the source IRAS 2A, which is rather a peculiar case and is considered here as an outlier. While similar correlations for the momentum flux of molecular flows are well known [5, 7], this is the first time that such a correlation is shown to hold also for the [OI] ejecta emission related to embedded protostars, and it provides possible indications that the mass-loss phenomena also traced in atomic lines are directly linked to the accretion processes occurring very close to protostars.

If we assume that the bona-fide ejecta from protostars along all evolutionary stages is atomic, as evidenced from the micro-jets of T-Tauri stars [1], then the observed [OI] emission can very well represent the same atomic gas. In such a scenario, the molecular outflow emission traced in the CO lines would correspond

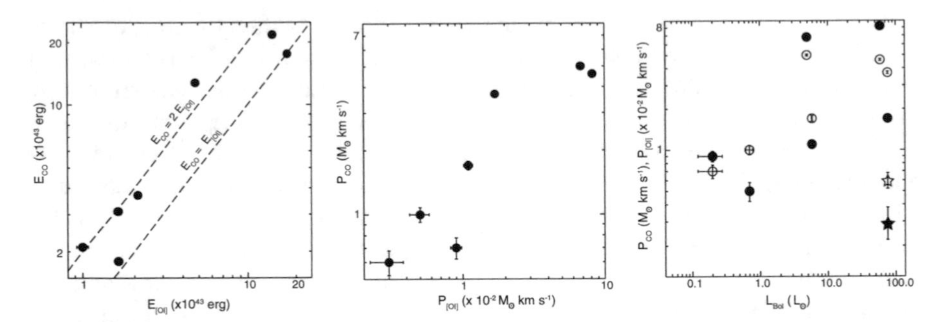

Fig. 2 Comparison of the energy and momentum (left and center panels, respectively) derived from the analysis of the [OI] emission with the corresponding CO values from [13], as listed in [8]. The CO and [OI] momenta (empty and filled circles, respectively) are plotted against the embedded-source bolometric luminosities in the right-most panel. In the same panel, filled and empty stars correspond to the east-west outflow of source IRAS 2A which is considered an outlier

to jet-entrained ambient gas in the surrounding environment of forming stars. While the physical structure and appearance of the molecular outflows would be dictated by the dominant modes of interaction between the underlying atomic jet with the surrounding medium [3], a causal link between the two flows would translate into conserving the momentum. Given that the [OI] velocity is $\sim$10 times higher than the CO gas, the [OI] momentum is $\sim$1% of the CO and that the total energy is equally distributed between the two flows, then we find that the CO appears to be 10^2–10^4 times more massive than the [OI] flow. In fact, the mass estimates in the previous paragraph support the higher mass ratio. To a first approach, these results suggest that the atomic jet does not possess enough momentum to drive the observed CO outflows. Alternatively, there may be significant [OI] reservoirs in the atomic flow where the gas is not excited and therefore not detected, an assumption that can hold in the context of [OI] excitation in shocks. In this scenario, the atomic gas is excited within a chain of shocks (internal working surfaces) and then cools radiatively in the post-shock zone. Therefore the cumulative size of the regions where the emission is arising would be very small compared to the total length of the flow [9]. In this case, the [OI] emission measures only the gas excited in shocks, but not the oblique atomic gas between shocks that is not radiating, but is still moving at relatively high velocities and therefore significantly contributes to stirring up a mixing turbulent layer which is observed in molecular line emission. In such a scenario the [OI] emission observed can indeed support the assumption that the protostellar ejecta are predominantly atomic.

In the discussion above we assumed that the atomic gas mainly represents ejecta from a protostar that is consequently excited in shocks. There is enough evidence, however, that the observed shocks are dissociative J-type, indicating that at least some of the atomic gas is produced in-situ from the dissociation of oxygen-bearing molecules such as CO and H_2O, at the location of the shocks. In this scenario, the oxygen production would strongly depend on the shock conditions such as the shock velocity and local density of the gas. Even though the shock conditions can have some affinity to the protostar, we would expect that the atomic [OI] production and excitation in shocks would show little correlation to the properties of the forming star, which is in contrast with the correlation shown in the right panel of Fig. 2. In that case, the observed correlation suggests that each flow is dynamically linked to the driving source. The relation found between the momentum and the energy of the atomic and molecular ejecta agrees with the predictions of the numerical simulations [10], which studied the formation and early evolution of protostars considering two-component (jet and outflow) ejecta in a nested configuration. In these simulations, the low-velocity outflow ($u_{out} < 50\,\mathrm{km\,s^{-1}}$) originates from the protostellar disk, is less collimated, and is rather constant during the main accretion phase. On the other hand, the high-velocity jet is highly variable because of episodic accretion events that can be created by a number of different instabilities in the accretion disk (e.g. gravitational instability [17], or a combination of gravitational instability and magnetic dissipation [10]). In this configuration, [10] demonstrate that the momentum generated by the episodic jet outbursts is always much lower than the more constant outflow, but the kinetic energy of the jet has spikes that exceed the energy of the outflows. As a result, the time-averaged values for the jet kinetic energy can account for a significant fraction of the outflow energy, but in contrast the jet momentum is too low to reach the outflow values even at the highest outburst peaks.

Acknowledgements This research was supported by the Austrian Research Promotion Agency (FFG) under the framework of the Austrian Space Applications Program (ASAP) projects JetPro* and PROTEUS (FFG-854025, FFG-866005).

References

1. Agra-Amboage, V., Dougados, C., Cabrit, S., Garcia, P. J. V., & Ferruit, P. 2009, A&A, 493, 1029
2. Arce, H. G., & Goodman, A. A. 2001, ApJ, 554, 132
3. Arce, H. G., Shepherd, D., Gueth, F., et al. 2007, Protostars and Planets V, 245
4. Bally, J., Reipurth, B., Lada, C. J., & Billawala, Y. 1999, AJ, 117, 410
5. Bontemps, S., Andre, P., Terebey, S., & Cabrit, S. 1996, A&A, 311, 858
6. Dionatos, O., Nisini, B., Garcia Lopez, R., et al. 2009, ApJ, 692, 1
7. Dionatos, O., Nisini, B., Codella, C., & Giannini, T. 2010b, A&A, 523, A29
8. Dionatos, O., & Güdel, M. 2017, A&A, 597, A64
9. Hollenbach, D., & McKee, C. F. 1989, ApJ, 342, 306
10. Machida, M. N. 2014, ApJ, 796, L17
11. Masson, C., R.; Chernin, L., M., 1993, ApJ, 414, 230
12. Nisini, B., Santangelo, G., Giannini, T., et al. 2015, ApJ, 801, 121
13. Plunkett, A. L., Arce, H. G., Corder, S. A., et al. 2013, ApJ, 774, 22
14. Podio, L., Kamp, I., Flower, D., et al. 2012, A&A, 545, A44
15. Raga, A. C., Noriega-Crespo, A., Carey, S. J., & Arce, H. G. 2013, AJ, 145, 28
16. Yu, K. C., Billawala, Y., & Bally, J. 1999, AJ, 118, 2940
17. Vorobyov, E. I., & Basu, S. 2006, ApJ, 650, 956

Part V
A Tribute to K. Tsinganos

A Short Tribute to Kanaris Tsinganos, Conclusions to This Book

J. J. G. Lima, C. Sauty, and N. Vlahakis

1 Introduction

This contribution is a special one dedicated to Kanaris Tsinganos, who promoted MHD Astrophysics and as such was a key leader in the JETSET network. This network finished in 2008 with a great international meeting in Rhodes organised by Kanaris. When this meeting was first organised, it was planned to have a special session to acknowledge Kanaris' contribution to this network and to science. This is a collection thoughts of three of his former students, now colleagues of him.

J. J. G. Lima (✉)
Instituto de Astrofísica e Ciências do Espaço, Universidade do Porto, Porto, Portugal

Faculdade de Ciências, Departamento de Física e Astronomia, Universidade do Porto, Porto, Portugal
e-mail: jlima@astro.up.pt

C. Sauty
Laboratoire Univers et Théories, Observatoire de Paris, Université Paris Diderot, Meudon, France
e-mail: Christophe.Sauty@obspm.fr

N. Vlahakis
Section of Astrophysics, Astronomy and Mechanics, Department of Physics, National and Kapodistrian University of Athens, Athens, Greece
e-mail: vlahakis@phys.uoa.gr

C. Sauty (ed.), *JET Simulations, Experiments, and Theory*,
Astrophysics and Space Science Proceedings 55,
https://doi.org/10.1007/978-3-030-14128-8_21

2 Kanaris Tsinganos, a Truly Inspiring Scientist, by J. J. G. Lima

Kanaris Tsinganos is a very influential figure in the area of Plasma Astrophysics, Magnetohydrodynamics and, in particular, in Solar and Stellar Winds, amongst other areas.

Our paths crossed in 1991 when he was a Visiting Professor at the University of St.Andrews (UK) where I was taking my Ph.D. My work involved a generalization of one of his models on "Topologies of 2D helicoidal hydrodynamic solutions". The timing was thus perfect for me to learn and be inspired by him. And that was the case right from our first meetings where I could witness and benefit form the way he puts very clearly his ideas.

In the following year, I was given the opportunity to spend three months at the University of Crete in Heraklion to work more closely under his supervision, now on a generalization of 2D helicoidal magnetohydrodynamic solutions. Kanaris Tsinganos offered me an invaluable orientation during this crucial period of my Ph.D. research. Him and his family welcomed me so kindly to Crete, making me feel at home during that period.

Later, in 1994, he was a Visiting Professor at the University of Porto where he lectured a graduate course in Plasma Astrophysics and we could collaborate to prepare a manuscript for publication on the application of the hydrodynamic model to the heliolatitudinal gradient of the solar wind during solar minimum conditions.

During those periods, I was very privileged to participate in several discussions involving him and some of his other students and fellows. I could witness his extremely sharp physical intuition, the joy he feels in sharing his ideas with others and in opening new horizons for students. Those moments were truly inspiring for all of us present. And they still are, each time we meet and discuss. I'm very fortunate to continue to collaborate with him until now.

The academic legacy of Kanaris Tsinganos can be testified by the many Ph.D. students and visitors during their Ph.D. who were supervised by him, some of them now distinguished international academicians, like Christophe Sauty (Full Professor at the Observatoire de Paris, France), Nektarios Vlahakis (Associate Professor at the University of Athens, Greece), Titos Matsakos (JDX Consulting UK), Odysseas Dionatos (Research Associate at the University of Vienna, Austria), just to name those that I personally know. Their scientific careers were inspired by the example of Kanaris Tsinganos and by his leading role in creating a cohesive team in the area which continues to collaborate in several topics until now.

Almost three decades since I first met Kanaris, this meeting celebrating 10 years of JETSET FP6 has given me the opportunity to see him once again, to discuss science with him and other friends and to express my gratitude. Thank you, Christophe, for organizing this workshop and the special session to acknowledge Kanaris' contribution to science. Thank you, Kanaris, for continuing to be an inspiration and example to all of us.

3 A Leading Scientist, a Colleague and a Friend, by Christophe Sauty

I want to acknowledge here the crucial role Kanaris has played in my career. We first met in 1988. I was lucky to do my Master internship in Crete under the supervision of Joseph Ventura and Nick Kylafis, thanks to Jean Heyvaerts. He provided me one of the first European grant, an Erasmus one I guess. I knew that Kanaris was working in the field of MHD astrophysics and I was very much interested. Once he invited me in his office and I asked him, if I could work under his supervision. Kanaris proposed me a PhD project. This is how a long standing collaboration started that with years turned into a real friendship. It started in fact as my civil service (cooperation), which was mandatory at that time in France and then officially into a real PhD. What a chance! The PhD proposal, he submitted to Jean Heyvaerts, "a quick proposal draft" is a six page long document (see Fig. 1). This is already saying how deeply he goes into the details and his commitment to science.

I am indebt to both Kanaris Tsinganos, as my practial and actual supervisor in Crete, and Jean Heyvaerts, as my official supervisor. There is thus a real filiation, spiritually meaning, with Evry Shatzmann, on one hand, and Gene Parker and Subrahmanyan Chandrashekar on the other hand. Great people I met and could discuss with thanks to Kanaris and Jean. I had opportunities to benefit from their advices and comments. Needless to say that Gene Parker has a great opinion of his

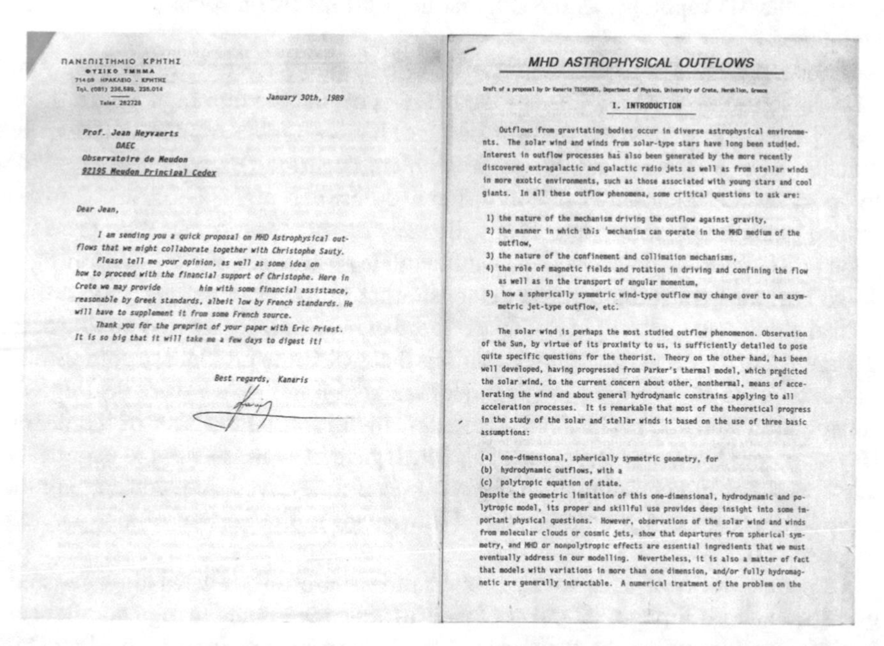

ΠΑΝΕΠΙΣΤΗΜΙΟ ΚΡΗΤΗΣ
ΦΥΣΙΚΟ ΤΜΗΜΑ
71409 ΗΡΑΚΛΕΙΟ - ΚΡΗΤΗΣ
Τηλ. (081) 236.589, 238.014
Telex 262728

January 30th, 1989

Prof. Jean Heyvaerts
DAEC
Observatoire de Meudon
92195 Meudon Principal Cedex

Dear Jean,

I am sending you a quick proposal on MHD Astrophysical outflows that we might collaborate together with Christophe Sauty.

Please tell me your opinion, as well as some idea on how to proceed with the financial support of Christophe. Here in Crete we may provide him with some financial assistance, reasonable by Greek standards, albeit low by French standards. He will have to supplement it from some French source.

Thank you for the preprint of your paper with Eric Priest. It is so big that it will take me a few days to digest it!

Best regards, Kanaris

MHD ASTROPHYSICAL OUTFLOWS

Draft of a proposal by Dr Kanaris TSINGANOS, Department of Physics, University of Crete, Heraklion, Greece

I. INTRODUCTION

Outflows from gravitating bodies occur in diverse astrophysical environments. The solar wind and winds from solar-type stars have long been studied. Interest in outflow processes has also been generated by the more recently discovered extragalactic and galactic radio jets as well as from stellar winds in more exotic environments, such as those associated with young stars and cool giants. In all these outflow phenomena, some critical questions to ask are:

1) the nature of the mechanism driving the outflow against gravity,
2) the manner in which this mechanism can operate in the MHD medium of the outflow,
3) the nature of the confinement and collimation mechanisms,
4) the role of magnetic fields and rotation in driving and confining the flow as well as in the transport of angular momentum,
5) how a spherically symmetric wind-type outflow may change over to an asymmetric jet-type outflow, etc.

The solar wind is perhaps the most studied outflow phenomenon. Observation of the Sun, by virtue of its proximity to us, is sufficiently detailed to pose quite specific questions for the theorist. Theory on the other hand, has been well developed, having progressed from Parker's thermal model, which predicted the solar wind, to the current concern about other, nonthermal, means of accelerating the wind and about general hydrodynamic constrains applying to all acceleration processes. It is remarkable that most of the theoretical progress in the study of the solar and stellar winds is based on the use of three basic assumptions:

(a) one-dimensional, spherically symmetric geometry, for
(b) hydrodynamic outflows, with a
(c) polytropic equation of state.

Despite the geometric limitation of this one-dimensional, hydrodynamic and polytropic model, its proper and skillful use provides deep insight into some important physical questions. However, observations of the solar wind and winds from molecular clouds or cosmic jets, show that departures from spherical symmetry, and MHD or nonpolytropic effects are essential ingredients that we must eventually address in our modelling. Nevertheless, it is also a matter of fact that models with variations in more than one dimension, and/or fully hydromagnetic are generally intractable. A numerical treatment of the problem on the

Fig. 1 The two first pages of the draft of a PhD proposal sent by Kanaris Tsinganos to Jean Heyvaerts in 1989

former PhD student, Kanaris Tsinganos. I must also say that this is also the reason why the whole volume is dedicated to Jean Heyvaerts who passed away too early.

From Parker's influence, I guess, Kanaris has always had that perfect skill to combine a theoretical and mathematical approach with a physical sense for empirical intuition. Besides, Kanaris is a really tough worker and together we pursued the same track for years, namely the development of self similar solutions, to go deeper and deeper in the same furrow. Yet the results are beyond expectation. We started from the Solar wind, moved to young stellar objects, jets and accretion, and now by extending to relativistic jets in Kerr metrics, we have also explored Active Galactic Nuclei and Gamma Ray Bursts. Of course this is only a small part of Kanaris activity who is also involved in solar physics, instruments, ESA, etc...

Just to focalize on a very specific aspect, let me recall a few breakthroughs, on which he was a leader, namely the study of the famous critical points. This is probably one of the most challenging difficulties of steady MHD flows. We have explored that these critical points in the 2,5D case were 3D and no longer 2D topologies. This was the occasion for me to build the first 3D topology with wires. I presented this piece of art to my first meeting in Crieff, where I give my first talk, as a PhD student, thanks to Kanaris. As a matter of fact, Kanaris has always been very supportive with his students, giving them unique opportunities to have access to the fantastic albeit terrifying scientific world. In that respect I also have nice memories, a few years later, of the conference in honour of L. Mestel in Cambridge that I attended again thanks to him. The meeting in May was an occasion to build a small copy of the 3D topology, as the original is lost a bit like a roman copy of greek antiquities.

Another breakthrough was to prove that critical points are not the usual slow and fast magnetosonic surfaces, besides the Alfvén one, but the limiting characteristics, see [2]. Something analogous to the difference between an ergosphere and the event horizon. As a matter of fact, the same problem occurred (Brandon Carter's PhD) to convince the community of the difference between ergospheres and horizons around black holes, in the late 1960s. Besides the publication in 1996, there was a long discussion again between us and Jean Heyvaerts, see Fig. 2. The left figure is the beginning of a three pages demonstration of the classical derivation of the critical points and on the right is Kanaris' answer, who emphasizes the mixing of the Bernoulli and the Grad-Shafranov equations that makes the problem so complicated. The end of the story came from Jean Heyvaerts saying that we were dealing with very complicated and sophisticated model. Coming from the author of the famous Heyvaerts and Norman, 1989, paper [1], which gave us so much work to understand, this is a rather big compliment. This is however a deserved one, considering that Kanaris never let things unfinished and always go to the bottom of any problem, exhausting it.

Of course the story does not stop now, retirement is only a word on paper, and I guess we have still plenty of years of fruitful research to pursue together. Because Kanaris has a lot of humor, which is not the least of his qualities, he qualified him self in the past my PhD as the old testament and my habilitation has the new one.

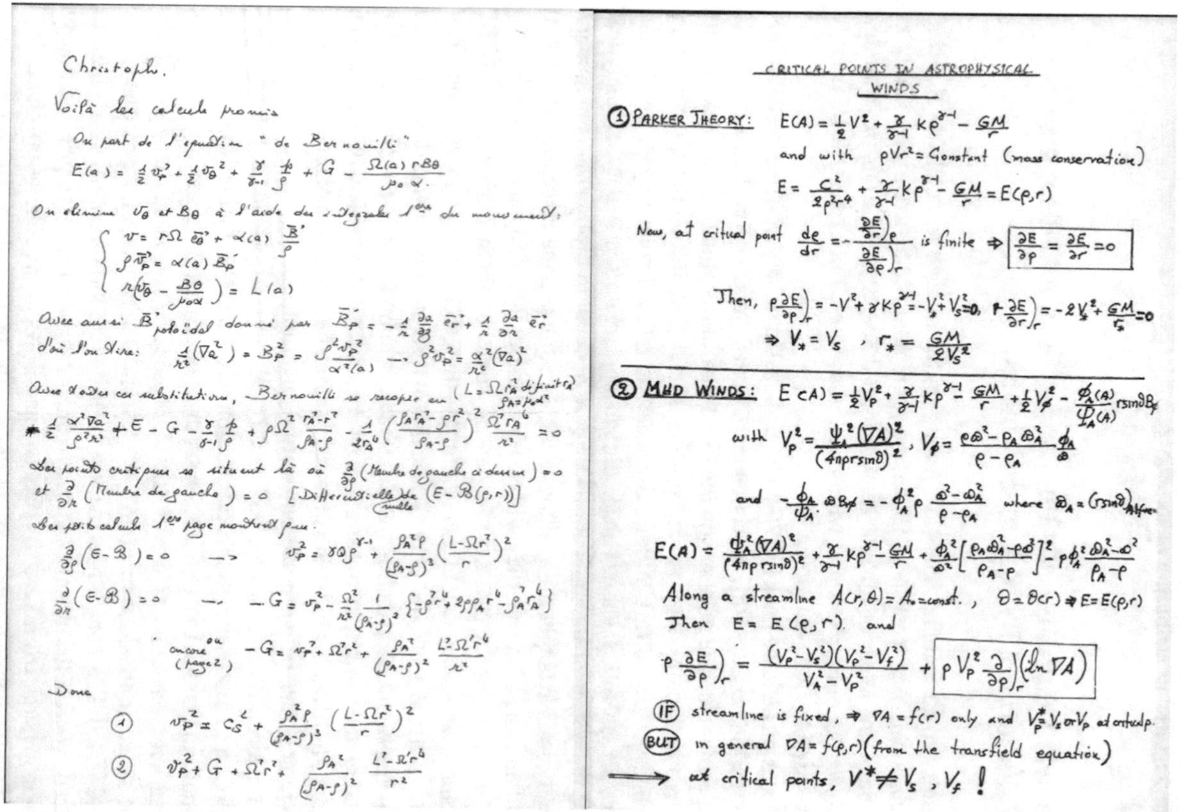

Christophe.

Voilà les calculs promis

On part de l'équation "de Bernoulli"

$$E(a) = \frac{1}{2}v_p^2 + \frac{1}{2}v_\theta^2 + \frac{\gamma}{\gamma-1}\frac{p}{\rho} + G - \frac{\Omega(a)\, r B_\theta}{\mu_0 \alpha}$$

On élimine v_θ et B_θ à l'aide des intégrales 1ère du mouvement:

$$\begin{cases} \vec{v} = r\Omega\,\vec{e}_\theta + \alpha(a)\dfrac{\vec{B}}{\rho} \\ \rho\vec{v}_p = \alpha(a)\vec{B}_p \\ r\left(v_\theta - \dfrac{B_\theta}{\mu_0\alpha}\right) = L(a) \end{cases}$$

Avec aussi $\vec{B}$ poloïdal donné par $\vec{B}_p = -\frac{1}{r}\frac{\partial a}{\partial z}\vec{e}_r + \frac{1}{r}\frac{\partial a}{\partial r}\vec{e}_z$

d'où l'on tire: $\frac{1}{r^2}(\nabla a)^2 = B_p^2 = \frac{\rho^2 v_p^2}{\alpha^2(a)} \rightarrow \rho^2 v_p^2 = \frac{\alpha^2}{r^2}(\nabla a)^2$

Avec toutes ces substitutions, Bernoulli se recopie en ($L = \Omega r_A^2$ définit r_A, $\rho_A = \mu_0\alpha^2$)

$$* \ \frac{1}{2}\frac{\alpha^2 \nabla a^2}{\rho^2 r^2} + E - G - \frac{\gamma}{\gamma-1}\frac{p}{\rho} + \rho\Omega^2\frac{r_A^2 - r^2}{\rho_A - \rho} - \frac{1}{2r_A^4}\left(\frac{\rho_A r_A^2 - \rho r^2}{\rho_A - \rho}\right)^2 \frac{\Omega^2 r_A^6}{r^2} = 0$$

Les points critiques se situent là où $\frac{\partial}{\partial\rho}$(Membre de gauche ci-dessus) $= 0$

et $\frac{\partial}{\partial r}$(Membre de gauche) $= 0$ [Différentielle de $(E - \mathcal{B}(\rho, r))$ nulle]

Les petits calculs 1ère page montrent que:

$$\frac{\partial}{\partial\rho}(E - \mathcal{B}) = 0 \longrightarrow v_p^2 = \gamma Q\rho^{\gamma-1} + \frac{\rho_A^2\rho}{(\rho_A-\rho)^3}\left(\frac{L - \Omega r^2}{r}\right)^2$$

$$\frac{\partial}{\partial r}(E - \mathcal{B}) = 0 \longrightarrow -G = v_p^2 - \frac{\Omega^2}{r^2}\frac{1}{(\rho_A-\rho)^2}\left\{-\rho^2 r^4 + 2\rho\rho_A r^4 - \rho_A r_A^4\right\}$$

encore ou (page 2) $-G = v_p^2 + \Omega^2 r^2 + \frac{\rho_A^2}{(\rho_A-\rho)^2}\frac{L^2\Omega^2 r^4}{r^2}$

Donc

① $v_p^2 = c_s^2 + \frac{\rho_A^2\rho}{(\rho_A-\rho)^3}\left(\frac{L - \Omega r^2}{r}\right)^2$

② $v_p^2 + G + \Omega^2 r^2 + \frac{\rho_A^2}{(\rho_A-\rho)^2}\frac{L^2 - \Omega^2 r^4}{r^2}$

CRITICAL POINTS IN ASTROPHYSICAL WINDS

① PARKER THEORY: $E(A) = \frac{1}{2}V^2 + \frac{\gamma}{\gamma-1}K\rho^{\gamma-1} - \frac{GM}{r}$

and with $\rho V r^2 =$ Constant (mass conservation)

$$E = \frac{C^2}{2\rho^2 r^4} + \frac{\gamma}{\gamma-1}K\rho^{\gamma-1} - \frac{GM}{r} = E(\rho, r)$$

Now, at critical point $\frac{d\rho}{dr} = -\frac{\left.\frac{\partial E}{\partial r}\right)_\rho}{\left.\frac{\partial E}{\partial \rho}\right)_r}$ is finite $\Rightarrow$ $\boxed{\frac{\partial E}{\partial \rho} = \frac{\partial E}{\partial r} = 0}$

Then, $\rho\left.\frac{\partial E}{\partial\rho}\right)_r = -V^2 + \gamma K\rho^{\gamma-1} = -V_*^2 + V_s^2 = 0$, $r\left.\frac{\partial E}{\partial r}\right)_\rho = -2V_*^2 + \frac{GM}{r_*} = 0$

$$\Rightarrow V_* = V_s, \quad r_* = \frac{GM}{2V_s^2}$$

② MHD WINDS: $E(A) = \frac{1}{2}V_p^2 + \frac{\gamma}{\gamma-1}K\rho^{\gamma-1} - \frac{GM}{r} + \frac{1}{2}V_\phi^2 - \frac{\Phi_A(A)}{\Psi_A(A)} r\sin\theta B_\phi$

with $V_p^2 = \frac{\Psi_A^2(\nabla A)^2}{(4\pi\rho r\sin\theta)^2}$, $V_\phi = \frac{\rho\varpi^2 - \rho_A\varpi_A^2}{\rho - \rho_A}\frac{\Phi_A}{\varpi}$

and $-\frac{\Phi_A}{\Psi_A}\varpi B_\phi = -\Phi_A^2\rho\frac{\varpi^2 - \varpi_A^2}{\rho - \rho_A}$ where $\varpi_A = (r\sin\theta)_{Alfvén}$

$$E(A) = \frac{\Psi_A^2(\nabla A)^2}{(4\pi\rho r\sin\theta)^2} + \frac{\gamma}{\gamma-1}K\rho^{\gamma-1} - \frac{GM}{r} + \frac{\Phi_A^2}{\varpi^2}\left[\frac{\rho_A\varpi_A^2 - \rho\varpi^2}{\rho_A - \rho}\right]^2 - \rho\Phi_A^2\frac{\varpi_A^2 - \varpi^2}{\rho_A - \rho}$$

Along a streamline $A(r,\theta) = A_0 =$ const., $\theta = \theta(r) \Rightarrow E = E(\rho, r)$

Then $E = E(\rho, r)$ and

$$\rho\left.\frac{\partial E}{\partial\rho}\right)_r = \frac{(V_p^2 - V_s^2)(V_p^2 - V_f^2)}{V_A^2 - V_p^2} + \boxed{\rho V_p^2 \left.\frac{\partial}{\partial\rho}\right)_r (\ln\nabla A)}$$

(IF) streamline is fixed, $\Rightarrow$ $\nabla A = f(r)$ only and $V_p^* = V_s$ or V_f at critical p.

(BUT) in general $\nabla A = f(\rho, r)$ (from the transfield equation)

$\longrightarrow$ at critical points, $V^* \neq V_s, V_f$!

Fig. 2 Jean's demonstration of the critical points (first page, left) and Kanaris' proof (right) that critical points are modified in a free geometry in answer to Jean Heyvaerts

My personal comment is guess who in the next generation wrote the Coran, Talmud and Das Kapital, all that thanks to Kanaris' contribution?

With that I conclude a short tribute to Kanaris, but of course there would be much more to say and I hope we shall have another meeting in the future to cover other aspects of Kanaris' contribution to science.

4 Kanaris Tsinganos: My PhD Supervisor, a Colleague, a Collaborator, by Nektarios Vlahakis

Many thanks to Christophe Sauty for giving us the chance to celebrate Kanaris' retirement during this workshop. Of course retirement for a strong personality as Kanaris does not mean at all that he will stop doing research. On the contrary, without teaching obligations and without administrative duties he will be freer to focus on Astrophysics even more.

The occasion took me back to 1994 when, at the University of Crete in Heraklion, I started working on my PhD thesis under Kanaris' supervision. I still remember the first problem that he gave me on the Grad-Shafranov equation, but also that he left me free to explore other related problems, and how influential his comments were whenever I faced difficulties. Obviously he had a strong impact on me and my career in general. I learned from him to appreciate the beauty of analytical work and the power of its application to many magnetohydrodynamic phenomena in Astrophysics. He taught me to enjoy doing research and the discussions with him were, still are, and I am sure they will continue to be, full of joy and inspiration.

Over the years we wrote several papers, some of them during the JETSET collaboration, and also taught many courses together, always in harmony and cooperation. Since I am the only Greek that writes in this tribute, I should also stress his important contributions to the Greek community of Astrophysicists. He has served in many areas, notably he was a very successful president of the National Observatory of Athens, sacrificing much of his time being in the not always pleasant position to negotiate between institutions and governments.

As a Physicist Kanaris has worked on a broad range of problems. For example, I remember him discussing the fragmentation of air bubbles that he was observing in an aquarium that led to a publication on the hydrodynamic instability of buoyant fields and later was connected with the emergence of magnetic flux tubes in the Sun! As an Astrophysicist of course we all know him since he was one of the leaders of JETSET (among others) network. He has important contributions and is a recognized expert in the magnetohydrodynamic description of Astrophysical plasma outflows in general, from YSO to AGN jets and flows in the solar atmosphere.

Of course his beloved subject of research is the meridionally self-similar model of magnetized flows. He and his collaborators/students started in the 1980s with simple radial hydrodynamic models and progressed to magnetized flows even in Kerr spacetime in a long series of ten papers; the eleventh is actually in preparation!

Fig. 3 Kanaris Tsinganos at the foot of the Great Coupole at Meudon

He strongly believes in the robustness of analytical solutions, and he is apparently absolutely right, as they have been proven very stable in many works in which they were chosen as initial states in numerical simulations.

I could say much more, but I keep them for the time that Kanaris really decides to retire. For now I would like to wish that he continues to be strong and energetic, to transfer his knowledge and approach to science to all of us for many more years (Fig. 3).

5 Conclusions

Thus, this book comes to an end. This book was a unique collection of reviews on different aspects developed through the years after the JETSET network finished. This is also a memory of the meeting "JETSET, ten years later, what is next?" held in Meudon, in May 2018, with the support of Paris Observatory. This is also a way to celebrate Kanaris Tsinganos involvement and impulse to the study of jets in young stellar objects. An impulse he will certainly pursue, helping us to create a new European Network after so many years.

Acknowledgements C. Sauty thanks the CIAS and the Observatoire de Paris for the financial support they provided to organize this conference.

References

1. Heyvaerts J., Norman C., ApJ, 347, 1055 (1989)
2. Tsinganos K., Sauty C., Surlantzis G., Trussoni E., Contopoulos J., MNRAS, 283, 811 (1996)

Printed in the United States
By Bookmasters